Multiplicative Inequalities of Carlson Type and Interpolation

Multiplicative Inequalities of Carlson Type and Interpolation

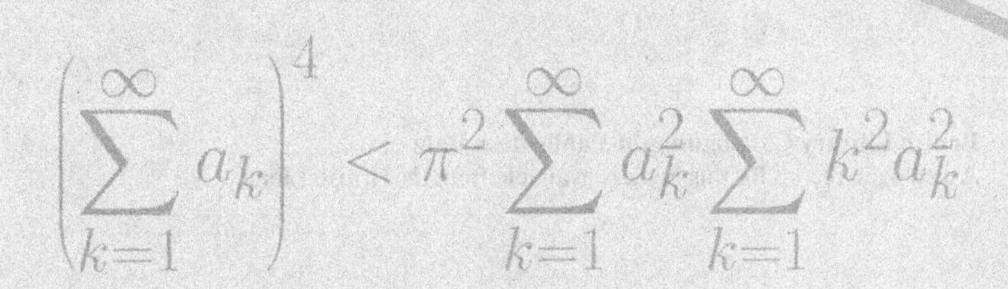

Leo Larsson
Uppsala University, Sweden

Lech Maligranda
Luleå University of Technology, Sweden

Josip Pečarić
University of Zagreb, Croatia

Lars-Erik Persson
Luleå University of Technology and Uppsala University, Sweden

World Scientific

NEW JERSEY • LONDON • SINGAPORE • BEIJING • SHANGHAI • HONG KONG • TAIPEI • CHENNAI

Published by

World Scientific Publishing Co. Pte. Ltd.
5 Toh Tuck Link, Singapore 596224
USA office: 27 Warren Street, Suite 401-402, Hackensack, NJ 07601
UK office: 57 Shelton Street, Covent Garden, London WC2H 9HE

Library of Congress Cataloging-in-Publication Data
Larsson, Leo, 1972–
Multiplicative inequalities of Carlson type and interpolation / Leo Larsson ... [et al.].
p. cm.
Includes bibliographical references and index
ISBN-13 978-981-256-708-6 (alk. paper)
ISBN-10 981-256-708-9 (alk. paper)
1. Inequalities (Mathematics) 2. Interpolation. 3. Numerical analysis. I. Title.

QA295.M86 2006
515'.26--dc22

2006042168

British Library Cataloguing-in-Publication Data
A catalogue record for this book is available from the British Library.

Printed in Singapore

Fritz David Carlson (1888–1952)

Preface

The following remarkable inequalities were stated and proved by the Swedish mathematician Fritz Carlson in 1934 (see [23]):

$$\text{(A)} \qquad \left(\sum_{k=1}^{\infty} a_k\right)^4 < \pi^2 \sum_{k=1}^{\infty} a_k^2 \sum_{k=1}^{\infty} k^2 a_k^2,$$

$$\text{(B)} \qquad \left(\int_0^{\infty} f(x)\,dx\right)^4 \leq \pi^2 \int_0^{\infty} f^2(x)\,dx \int_0^{\infty} x^2 f^2(x)\,dx.$$

These inequalities have later on attracted a great interest and they have been generalized and applied in an almost unbelievable way. In this book, we present and discuss some of the most important steps and results in this development. An inequality of the form

$$\|f\|_X \leq C \|f\|_{A_0}^{1-\theta} \|f\|_{A_1}^{\theta},$$

where X, A_0 and A_1 are normed spaces and $0 < \theta < 1$, or

$$\|f\|_X \leq C \prod_{j=1}^{m} \|f\|_{A_j}^{\theta_j},$$

where X and A_j are normed spaces, $0 < \theta_j < 1$, $j = 1, \ldots, m$ and $\sum_{j=1}^{m} \theta_j = 1$, will be referred to as a *Carlson type inequality*.

Obviously, this development has strongly contributed not only to the knowledge in the theory of inequalities, but also to the development and applications in other branches of pure mathematics (e.g. interpolation theory, harmonic analysis, theory of function spaces, etc.) as well as in applied mathematics (e.g. optimal sampling in signal processing) and history of mathematics.

In Chapter 0, we give a short description of some of the applications mentioned above, and some notation is established.

In Chapter 1, we first give a sketch of Carlson's rather complicated proof of (A), followed by some comments. We have also included two proofs due to G. H. Hardy, as well as some proofs based on different ideas. The chapter is concluded by a discussion of the case when the infinite series in (A) are replaced by finite sums.

In Chapter 2, we present and discuss a number of early extensions and complements of the discrete inequality (A) (e.g. those by Gabriel, Levin, Caton, Bellman, Landau, Levin–Stečkin, Levin–Godunova). Moreover, a very new extension and unification of the Landau and Levin–Stečkin results are stated and proved. Also some new, extended results concerning the case with finite sums are included.

In Chapter 3 we handle the corresponding question concerning the continuous inequality (B). In particular, some early results by Beurling, Kjellberg, Bellman, Sz. Nagy, and Yang–Fang are presented and discussed. Also, some new results concerning the case when $(0, \infty)$ is replaced by a bounded interval $(0, m)$ or (m^{-1}, m) are presented and proved.

In Chapter 4 we present, prove, and discuss a very crucial extension of Carlson's inequality (B) from 1948 by Levin.

A number of multi-dimensional variations and generalizations of Carlson's inequality (B) can be found in Chapter 5. The most crucial result for the further discussions is an extension of (B) to the case with general cones in $\mathbb{R}^n$ from 1998 by Barza–Burenkov–Pečarić–Persson. Also generalizations by Andrianov, Pigolkin, Bertolo–Fernandez, Kamaly and Kjellberg are presented and discussed. Moreover, a newer generalization of Kamaly's result is included.

In Chapter 6, a very recent Carlson type inequality for weighted Lebesgue spaces on general measure spaces by Larsson is presented and proved. Also the more complicated case with two or more factors in product measure spaces is treated. We pronounce that general results from real interpolation theory are used in the proof in a crucial way.

In Chapter 7, we discuss further the close connection between real interpolation theory and Carlson type inequalities. In order to make the text self-contained, some elementary definitions and facts from real interpolation theory are included. Moreover, it is pointed out how Carlson type inequalities can be used to prove well-known and new embeddings between some function spaces.

As has been seen in previous chapters, real interpolation theory can be used to prove some new Carlson type inequalities. In Chapter 8, we will discuss the reverse relation, namely that Carlson type inequalities can imply new results in interpolation theory (e.g. concerning the Gustavsson–Peetre $\pm$ method). In particular, a fairly new (optimal) block Carlson type inequality by Kruglyak–Maligranda–Persson is proved and discussed. Moreover, it is pointed out how this inequality can be used to (positively) answer a fairly old question raised by Peetre concerning the interpolation functor $\langle\cdot\rangle_\varphi$ named after him.

A number of further results, applications and open questions have been collected in Chapter 9.

A historical note on Fritz David Carlson and his work can be found in Appendix A. In particular, we remark that this unique information has been complemented by his son, Professor Per Carlson (Royal Institute of Technology, Stockholm, Sweden).

Finally, in Appendix B, we have included for the reader's convenience, a translation from the French of the original Carlson paper [23].

The starting point for this book was the Ph. D. thesis of the first named author, in which the historical development of the inequalities (A) and (B) was briefly described. Our motivation for writing this book is not only to fill in the obvious gap that there exists no textbook on this subject, but also to collect all present knowledge in such a unified form that it

(a) clearly shows the development of this fascinating theory in a historical perspective,
(b) clearly points out the close relations between various parts of the theory, and also to other areas within mathematics and some applications,
(c) clarifies the close relation between Carlson type inequalities and interpolation theory,
(d) can be used directly as a source book for further research of different kinds (e.g. a number of open questions are raised).

We strongly believe that these facts will imply that this book can attract a fairly broad audience of readers from different fields within mathematics, but also in applied sciences and history.

The Intended Audience of this Book

This book contains almost all results we know concerning inequalities of Carlson type and also all techniques of proof, as well as a number of open questions, so it should be a source book for everybody interested in or working with the theory of inequalities. Furthermore, it is of interest for mathematicians working in several disciplines within analysis and its applications (e.g. interpolation theory, harmonic analysis, function spaces, etc.). It is also of interest in applications outside the mathematical area (e.g. within optimal control in signal processing). Finally, it is of value for anybody interested in the history of mathematics (e.g. because it contains a fairly complete historical development of what is today called Carlson type inequalities and also a historical note on Fritz David Carlson himself).

Acknowledgements

First of all, the authors would like to thank Professor Per Carlson, Stockholm, for providing us with some valuable information, which has improved our historical note on (his father) Fritz Carlson. He has also kindly provided us with the photography of Fritz Carlson. We would also like to thank Professor Jaak Peetre, Lund, for some generous advice and explanations concerning the early history of topics related to his interpolation method (see Chapters 8 and 9).

Luleå, Zagreb, Uppsala; January 2006

Leo Larsson, Lech Maligranda, Josip Pečarić, Lars-Erik Persson

Contents

Chapter 0

Introduction and Notation

This book is concerned with the following remarkable inequalities of F. Carlson [23]:

$$\text{(A)} \qquad \left(\sum_{k=1}^{\infty} a_k\right)^4 < \pi^2 \sum_{k=1}^{\infty} a_k^2 \sum_{k=1}^{\infty} k^2 a_k^2,$$

$$\text{(B)} \qquad \left(\int_0^{\infty} f(x)\,dx\right)^4 \leq \pi^2 \int_0^{\infty} f^2(x)\,dx \int_0^{\infty} x^2 f^2(x)\,dx.$$

We strongly believe that Carlson himself did not realize in 1934, when he formulated and proved the inequalities (A) and (B), that his inequalities later on would attract such an interest as can be seen from the descriptions and references presented in this book. This development has contributed to the general theory of inequalities, but also to the development and applications to a number of other areas within pure mathematics (e.g. interpolation theory, Fourier analysis, etc.), applied mathematics (e.g. optimal sampling problems within signal processing), as well as the history of mathematics (see e.g. several historical remarks concerning the development of (A) and (B) throughout this book and also the historical note on Fritz David Carlson (1888–1952) presented in Appendix A).

The inequalities (A) and (B), and their many generalizations and variations, turn out to have applications in various branches of mathematics. Here, we briefly mention some examples of this, which are covered in more detail inside the book.

A. Beurling [15] proved a version of the inequality (B), where the interval of integration is replaced by the whole real line. His version of the inequality was used in connection with the study of Fourier transforms of functions on $\mathbb{R}$. In fact, by using the Parseval identity, (B) can be thought of as an inequality estimating from above the integral of a function f by L_2-norms of the Fourier transform and the derivative of the Fourier transform of f.

A. Kamaly [40] has generalized the viewpoint of Beurling, thus regarding (B) as an inequality in Fourier analysis. He proved a new, multi-dimensional version of the inequality, and applied it to prove a sharp version of the Hausdorff–Young inequality for functions with small supports.

J. Bergh [11] applied the Beurling version of (B) to a problem in optimal sampling theory, where one wants to minimize a certain *pulse energy* given a maximal *error energy*. He was able to use the sharpness of the inequality to find out how densely we need to sample a smooth function with certain growth conditions in order to be able to reconstruct the function.

B. Kjellberg [43, 44] thought of the inequality (B) as stating that if the *moments* of the function f on the right-hand side are finite, then f is integrable. He gave necessary and sufficient conditions for certain moments, the finiteness of which should imply the integrability of the function.

J. Peetre [72] used a generalized version of (B), in its final form proved by V. I. Levin [56], in order to reduce the number of parameters involved when constructing certain interpolation spaces between Banach spaces, so called *espaces de moyennes*. Initially, those spaces were constructed using three parameters θ, p_0, and p_1. He applied the inequality as a step in the proof that the parameters p_0 and p_1 could be replaced by the single parameter p, defined by the relation

$$\frac{1}{p} = \frac{1-\theta}{p_0} + \frac{\theta}{p_1}.$$

In addition, the series on the right-hand side of a later version of (A) can be replaced by certain *block sums*, in this way improving some results regarding the Peetre $\langle\cdot\rangle_\varphi$ interpolation functor, in the sense that a larger class of φ can be used. This was done by N. Ya. Kruglyak, L. Maligranda, and L.-E. Persson [47].

By generalizing the inequalities (A) and (B) to hold for functions defined on general measure spaces, one can impose conditions on weight functions which give embeddings of real interpolation spaces into weighted Lebesgue spaces (see L. Larsson [50]). Also, by using recent versions of (B), it is possible to prove sharp, multi-dimensional embedding theorems of Sobolev type (see S. Barza, V. Burenkov, J. Pečarić, and L.-E. Persson [5]).

Finally, let us mention that the books by D. S. Mitrinovic [66], D. S. Mitrinovic, J. E. Pečarić, and A. M. Fink [67], and E. F. Beckenbach and

R. Bellman [8], all have separate sections devoted to (early versions of) Carlson type inequalities.

0.1 Notational Conventions

Below, we establish some notational conventions used in this book.

0.1.1 *Indices and Exponents*

We reserve the letter n for dimension in this book, while m is used as the upper limit in a finite sum or product. The letter k is used as the index in an infinite sum. The letters i and j are indices from a finite set, typically $\{0, 1\}$ or $\{1, \ldots, m\}$. θ and θ_i are always assumed to be numbers between 0 and 1. We use $1 - \theta$ and θ as the exponents in an expression like

$$\|f\|_{A_0}^{1-\theta} \|f\|_{A_1}^{\theta},$$

where two factors are involved. If an arbitrary finite number m of factors are used, as in

$$\prod_{i=1}^{m} \|f\|_{A_i}^{\theta_i},$$

we are assuming that

$$\sum_{i=1}^{m} \theta_i = 1.$$

We use prime notation to denote conjugate exponents. Thus, if $p > 1$, then p' is the number satisfying

$$\frac{1}{p} + \frac{1}{p'} = 1.$$

If $p = 1$, then $p' = \infty$ and conversely. This also applies to other letters than p.

0.1.2 *Constants*

The letter C is used in various contexts to denote an unspecified constant. Although the value of C will not change within a sequence of inequalities, C may have different meanings in different contexts. When a constant changes in a sequence of inequalities, we use prime notation; C', C'', ... or indices; C_1, C_2, ...

0.1.3 *Measure Spaces and Related Spaces*

The symbol (Ω, μ) is used to denote a measure space, consisting of a set Ω and a σ-finite measure μ. In this book, σ-algebras are either understood or not of importance.

0.1.3.1 *Lebesgue Spaces*

If $0 < p < \infty$, the Lebesgue space $L_p(\Omega, \mu)$ is the space of (complex-valued) measurable functions f for which

$$\int_\Omega |f|^p \, d\mu < \infty.$$

On $L_p(\Omega, \mu)$, the (quasi-)norm[1] is defined by

$$\|f\|_{L_p(\Omega,\mu)} = \left(\int_\Omega |f|^p \, d\mu \right)^{1/p}.$$

The space $L_\infty(\Omega, \mu)$ is defined as the space of measurable functions which are essentially bounded, and

$$\|f\|_{L_\infty(\Omega,\mu)} = \text{ess sup } |f|.$$

With identification of functions that are equal μ-almost everywhere, $L_p(\Omega, \mu)$ is a Banach space if $1 \leq p \leq \infty$.

When the measure space under consideration is clear from the context, we write $\|\cdot\|_p$ for the (quasi-)norm on $L_p(\Omega, \mu)$, $0 < p \leq \infty$.

Apart from the letter p as the exponents of Lebesgue spaces, we also use the indexed version p_0, p_1 etc., as well as other letters, with or without indices. This applies also to the weighted Lebesgue spaces defined below.

In the special case where the underlying space is $\mathbb{R}^n_+$ for some n, and the homogeneous measure is used, we will sometimes use an asterisk to denote the corresponding Lebesgue spaces. Thus $L^*_p(\mathbb{R}^n_+)$ consists of those measurable functions f on $\mathbb{R}^n_+$ such that

$$\|f\|^p_{L^*_p(\mathbb{R}^n_+)} = \int_{\mathbb{R}^n_+} |f(x)|^p \frac{dx}{|x|^n} < \infty.$$

---

[1] As is well known, this is a norm only if $p \geq 1$.

0.1.3.2 *Weighted Lebesgue Spaces*

Lebesgue spaces with weights are frequently used in this book. We adapt to the following convention.

Suppose that $0 < p < \infty$ and that w is a positive, measurable function on Ω. We denote by $L_p(\Omega, w^p\mu)$ the weighted Lebesgue space consisting of those f for which

$$\int_\Omega |fw|^p \, d\mu < \infty.$$

Usually, we do not use a special symbol for the norm on a weighted Lebesgue space, but write out the weight w in the unweighted norm:

$$\|fw\|_{L_p(\Omega,\mu)} .$$

0.1.4 *Interpolation Spaces*

The symbol $\bar{A}$ is the generic notation for a compatible couple (A_0, A_1) of normed spaces. This means that both A_0 and A_1 are continuously embedded in some universal Hausdorff topological vector space (this is needed e.g. in order to form the sum of the spaces A_0 and A_1). This notation applies to other letters than A. Thus, for instance, $\bar{X}$ is short for (X_0, X_1). Here, of course, we have to make clear what the spaces A_i are in order to define $\bar{A}$.

Given the couple $\bar{A}$ and the parameters θ and p, $\bar{A}_{\theta,p}$ denotes the space arising from the real interpolation method applied to the spaces A_0 and A_1 (see Chapter 7). In connection with the real interpolation method, the Peetre J and K functionals are defined. The letters J and K are reserved to denote these functionals.

0.1.5 *Linear Mappings Between Normed Spaces*

Suppose that X and Y are normed spaces. Then we will write $T : X \to Y$ to denote that T is a *bounded linear mapping* from X to Y. If T is linear, then T is bounded if and only if T is continuous.

If, in particular, X is a subspace of Y, then $X \subseteq Y$ will mean that $I : X \to Y$, where I is the identity mapping. In other words, the inclusion is *continuous*. Moreover, if c is a positive number, the symbol $X \overset{c}{\subseteq} Y$ will be used to indicate that the norm of the embedding $I : X \to Y$ does not exceed c.

If T is defined on X, and Z is a subspace of X, $T|_Z$ denotes the restriction of T to Z.

0.1.6 *Other*

The letter B will be used to denote the Beta function, defined by

$$B(\alpha,\beta) = \int_0^1 (1-s)^{\alpha-1}s^{\beta-1}\,ds = \int_0^1 (1-s)^{\alpha}s^{\beta}\frac{ds}{(1-s)s}, \quad \alpha,\beta>0.$$

The Beta function obeys the following easily verified symmetric and additive properties.

(1) $B(\alpha,\beta) = B(\beta,\alpha)$.
(2) $B(\alpha+1,\beta) = \dfrac{\alpha}{\alpha+\beta}B(\alpha,\beta)$.

Furthermore, if Γ denotes the Gamma function, defined by

$$\Gamma(z) = \int_0^\infty t^{z-1}e^{-t}\,dt = \int_0^\infty t^z e^{-t}\frac{dt}{t},$$

then we have the identity

$$B(\alpha,\beta) = \frac{\Gamma(\alpha)\Gamma(\beta)}{\Gamma(\alpha+\beta)}. \tag{0.1}$$

To see this, note that by switching to polar coordinates (ρ,φ) and then substituting r for $\rho(\cos\varphi+\sin\varphi)$, we can write

$$\begin{aligned}
\Gamma(\alpha)\Gamma(\beta) &= \int_0^\infty s^\alpha e^{-s}\frac{ds}{s}\int_0^\infty t^\beta e^{-t}\frac{dt}{t}\\
&= \int_0^\infty\int_0^\infty s^{\alpha-1}t^{\beta-1}e^{-(\alpha+\beta)}\,ds\,dt\\
&= \int_0^\infty\int_0^{\pi/2} (\rho\cos\varphi)^{\alpha-1}(\rho\sin\varphi)^{\beta-1}e^{-\rho(\cos\varphi+\sin\varphi)}\rho\,d\varphi\,d\rho\\
&= \int_0^\infty r^{\alpha+\beta}e^{-r}\frac{dr}{r}\int_0^{\pi/2}(\cos\varphi+\sin\varphi)^{-(\alpha+\beta)}(\cos\varphi)^{\alpha-1}(\sin\varphi)^{\beta-1}\,d\varphi\\
&= \Gamma(\alpha+\beta)\int_0^{\pi/2}\left(\frac{\cos\varphi}{\cos\varphi+\sin\varphi}\right)^{\alpha-1}\left(\frac{\sin\varphi}{\cos\varphi+\sin\varphi}\right)^{\beta-1}\frac{d\varphi}{(\cos\varphi+\sin\varphi)^2}
\end{aligned}$$

In the last integral, put

$$s = \frac{\sin\varphi}{\cos\varphi+\sin\varphi}.$$

Then

$$\frac{\cos\varphi}{\cos\varphi + \sin\varphi} = 1 - s$$

and

$$\frac{d\varphi}{(\cos\varphi + \sin\varphi)^2} = ds,$$

and the interval of integration transforms to $(0, 1)$. Thus

$$\begin{aligned}\Gamma(\alpha)\Gamma(\beta) &= \Gamma(\alpha+\beta)\int_0^1 (1-s)^{\alpha-1}s^{\beta-1}\,ds \\ &= \Gamma(\alpha+\beta)B(\alpha,\beta),\end{aligned}$$

which is (0.1).

Chapter 1

Carlson's Inequalities

This book is built around various versions and applications of the following result, first proved by F. Carlson [23].

Theorem 1.1 (Carlson, 1935) If $a_1, a_2, \ldots$ are real numbers, not all zero, then

$$\left(\sum_{k=1}^{\infty} a_k\right)^4 < \pi^2 \sum_{k=1}^{\infty} a_k^2 \sum_{k=1}^{\infty} k^2 a_k^2. \tag{1.1}$$

The constant π^2 is sharp.

Remark 1.1 Carlson's paper [23] was printed already in 1934, although the volume in which it was published did not appear until 1935. This may shed some light on the confusing fact that both 1934 and 1935 are seen in the literature.

Remark 1.2 The statement that the constant is sharp means that if it is replaced by any smaller number, then there is a sequence of numbers a_k for which the inequality is not true. A similar remark applies throughout this book whenever sharp, or best, constants are mentioned.

In this opening chapter, we first give a brief sketch of Carlson's own proof of Theorem 1.1 in Section 1.1. In Section 1.2, we present another way of proving the inequality (1.1), by a clever trick due to G. H. Hardy [32]. It turns out that his method can be extended to prove far more general versions of the inequality. We also recover a second proof of Hardy, based on some knowledge of Fourier series. Section 1.3 contains some other proofs of Carlson's inequality, and in Section 1.4, we investigate the case where the infinite series in (1.1) are replaced by finite sums. In Appendix B, a translation from the French of Carlson's paper [23] is given.

1.1 Carlson's Proof

In this section, we briefly sketch Carlson's proof of Theorem 1.1, which is based on the theory of analytic functions. Carlson realized that it is not possible to use the Hölder–Rogers inequality[1] directly to prove (1.1), by arguing as follows: For any $h > 1$, we can write

$$a_k = k^{-\frac{h}{2}} a_k^{\frac{1}{2}} (k^h a_k)^{\frac{1}{2}},$$

and so, by the Hölder–Rogers inequality with exponents 2, 4, 4, it follows that

$$\begin{aligned}\left(\sum_{k=1}^{\infty} a_k\right)^4 &= \left(\sum_{k=1}^{\infty} k^{-\frac{h}{2}} a_k^{\frac{1}{2}} (k^h a_k)^{\frac{1}{2}}\right)^4 \\ &\leq \left(\sum_{k=1}^{\infty} k^{-h}\right)^2 \sum_{k=1}^{\infty} a_k^2 \sum_{k=1}^{\infty} k^2 a_k^2 \\ &= C(h) \sum_{k=1}^{\infty} a_k^2 \sum_{k=1}^{\infty} k^2 a_k^2,\end{aligned}$$

where

$$C(h) = \left(\sum_{k=1}^{\infty} k^{-h}\right)^2.$$

To get (1.1), one would like to put $h = 1$ in this inequality. However,

$$\lim_{h \searrow 1} C(h) = \infty.$$

It therefore seems as Carlson at this moment regarded his inequality (1.1) as a limiting case which is not covered by the Hölder–Rogers inequality. Hence, it must have been a big surprise for him when G. H. Hardy presented a proof of (1.1) based merely on the Schwarz inequality (see Section 1.2 below).

[1]This is what is usually called Hölder's inequality. However, for historical reasons (see [62]), we choose to call it the Hölder–Rogers inequality.

First of all, Carlson showed that (1.1) holds for *some* constant C (in place of π^2), by using the following argument. Let

$$f(z) = \sum_{k=1}^{\infty} a_k z^k.$$

Then, since $f(0) = 0$, we have

$$\begin{aligned} f(1)^2 &= 2\int_0^1 f(z)f'(z)\,dz \\ &< 2\int_{-1}^1 |f(z)f'(z)|\,dz \\ &\le \int_0^{2\pi} |f(e^{i\varphi})f'(e^{i\varphi})|\,d\varphi \end{aligned}$$

and hence the Schwarz inequality yields

$$\begin{aligned} \left(\sum_{k=1}^{\infty} a_k\right)^4 = f(1)^4 &< \int_0^{2\pi} |f(e^{i\varphi})|^2\,d\varphi \int_0^{2\pi} |f'(e^{i\varphi})|^2\,d\varphi \\ &= 2\pi\sum_{k=1}^{\infty} a_k^2 \cdot 2\pi\sum_{k=1}^{\infty} k^2 a_k^2. \end{aligned}$$

This is (1.1) with π^2 replaced by $4\pi^2$.

Consider next finitely many non-negative numbers a_k, $k = 1, \dots, m$. We normalize so that

$$\sum_{k=1}^{m} a_k = 1.$$

Let

$$R = \sum_{k=1}^{m} a_k^2.$$

We then seek the minimum of the sum

$$S = \sum_{k=1}^{m} k^2 a_k^2$$

subject to these constraints. A Lagrange multiplier argument then implies that for some real numbers λ and μ, we have

$$a_k = \frac{\lambda}{k^2 + \mu}, \quad k = 1, \dots, m.$$

Therefore, the problem is reduced to finding the minimum of σ in the relation

$$\sum_{k=1}^{m} \frac{1}{(k^2+\mu)^2} \sum_{k=1}^{m} \frac{k^2}{(k^2+\mu)^2} - \sigma \left(\sum_{k=1}^{m} \frac{1}{k^2+\mu} \right)^4 = 0.$$

After a change of variables and a variational argument, Carlson then concludes that the sought constant C is the reciprocal value of the minimum of the function ω of z, implicitly defined by

$$\sum_{k=1}^{\infty} \left(\frac{2z^2}{k^2-z^2} \right)^2 \sum_{k=1}^{\infty} k^2 \left(\frac{2z^2}{k^2-z^2} \right)^2 - \omega \left(\sum_{k=1}^{\infty} \frac{2z^2}{k^2-z^2} \right)^4 = 0, \qquad (1.2)$$

where z is a real or purely imaginary number satisfying $z^2 < 1$. After substituting $\pi z = u$, these infinite series can be evaluated:

$$\begin{aligned} \sum_{k=1}^{\infty} \frac{2z^2}{k^2-z^2} &= 1 - u \cot u \\ &= \frac{\sin u - u \cos u}{\sin u}, \end{aligned}$$

$$\begin{aligned} \sum_{k=1}^{\infty} \left(\frac{2z^2}{k^2-z^2} \right)^2 &= u^2 + u^2 \cot^2 u + u \cot u - 2 \\ &= \frac{u^2 + u \sin u \cos u - 2 \sin^2 u}{\sin^2 u}, \end{aligned}$$

and

$$\begin{aligned} \sum_{k=1}^{\infty} k^2 \left(\frac{2z^2}{k^2-z^2} \right)^2 &= \frac{1}{\pi^2} (u^4 + u^4 \cot^2 u - u^3 \cot u) \\ &= \frac{1}{\pi^2} \frac{u^4 - u^3 \sin u \cos u}{\sin^2 u}. \end{aligned}$$

By multiplying through by $\pi^2 \sin^4 u$, (1.2) thus transforms to

$$\pi^2 \omega (u \cos u - \sin u)^4 = u^3 (u - \sin u \cos u)(u^2 + u \sin u \cos u - 2 \sin^2 u).$$

For $u \in (-\pi, \pi)$ or u purely imaginary, it follows from this that $\pi^2 \omega > 1$ and $\pi^2 \omega \to 1$ as u tends to infinity along the imaginary axis. Thus the non-strict inequality (1.1) holds, and we can also conclude that for any $\epsilon > 0$, there is a sequence a_k for which

$$\left(\sum_{k=1}^{\infty} a_k\right)^4 > (\pi^2 - \epsilon) \sum_{k=1}^{\infty} a_k^2 \sum_{k=1}^{\infty} k^2 a_k^2.$$

Hence the constant $C = \pi^2$ is sharp. Finally, to show that we in fact have strict inequality in (1.1) for non-zero sequences, Carlson applies the *continuous* version of the inequality

$$\left(\int_0^{\infty} f(x)\,dx\right)^4 \leq \pi^2 \int_0^{\infty} f^2(x)\,dx \int_0^{\infty} x^2 f^2(x)\,dx \tag{1.3}$$

(to be discussed in Chapter 3) to the function

$$f(x) = e^{-\frac{x}{2}} \sum_{k=0}^{\infty} (-1)^k a_{k+1} \rho^k L_k(x),$$

where L_k denotes the kth Laguerre polynomial and ρ is a real number < 1. With this choice of f, (1.3) becomes

$$\left(\sum_{k=1}^{\infty} a_k\right)^4 \leq \frac{\pi^2}{16} \sum_{k=1}^{\infty} a_k^2 \sum_{k=1}^{\infty} [(6k^2 - 6k + 2)a_k^2$$
$$+8(k-1)^2 a_{k-1} a_k + 2(k-1)(k-2) a_{k-2} a_k],$$

and the last sum on the right-hand side here is strictly dominated by

$$16 \sum_{k=1}^{\infty} \left(\left(k - \frac{1}{2}\right)^2 + \frac{3}{16}\right) a_k^2.$$

This shows that equality can never occur in (1.1).

Remark 1.3 Inspired by Carlson's proof above, we show directly that the constant π^2 cannot be replaced by a smaller constant. Let $\lambda > 0$, and put

$$a_k = \frac{\lambda}{k^2 + \lambda^2}, \quad k = 1, 2, \ldots.$$

We may then estimate the infinite series in (1.1) by integrals. We have

$$\sum_{k=1}^{\infty} a_k \geq \int_0^{\infty} \frac{\lambda\,dx}{x^2 + \lambda^2} - \frac{1}{\lambda} = \frac{\pi}{2} - \frac{1}{\lambda},$$

$$\sum_{k=1}^{\infty} a_k^2 \leq \int_0^{\infty} \frac{\lambda^2\, dx}{(x^2+\lambda^2)^2} = \frac{\pi}{4\lambda},$$

and

$$\sum_{k=1}^{\infty} k^2 a_k^2 \leq \int_0^{\infty} \frac{\lambda^2 x^2\, dx}{(x^2+\lambda^2)^2} = \frac{\lambda\pi}{4}.$$

Thus any constant C appearing in place of π^2 in (1.1) is bounded below by

$$\left(\frac{\pi}{4} - \frac{1}{\lambda}\right)^4 \Big/ \frac{\pi}{4\lambda}\frac{\lambda\pi}{4}.$$

Letting $\lambda \to \infty$ yields $C \geq \pi^2$.

Remark 1.4 The inequality (1.1) may also be written

$$\left(\sum_{k=1}^{\infty} |a_k|\right)^4 < \pi^2 \sum_{k=1}^{\infty} |a_k|^2 \sum_{k=1}^{\infty} k^2 |a_k|^2,$$

and it is then true also for complex sequences $\{a_k\}_{k=1}^{\infty}$. A similar remark holds true for other inequalities in this book, although we have chosen to state the result in their original formulations, thus sometimes refraining from mentioning the obvious possibility of using complex sequences.

Remark 1.5 We conclude this section by emphasizing that the continuous version (1.3) of Carlson's inequality and its many generalizations are also central in this book. We will explore them in greater detail later, starting in Chapter 3.

1.2 Hardy's Proofs

In 1936, G. H. Hardy [32] published two elementary proofs of Theorem 1.1. Contrary to Carlson's belief, described in the previous section, the following elementary proof shows that only the Schwarz inequality is needed to prove (1.1), provided that a clever trick is used.

Proof 1 (Hardy, 1936). Let α and β be positive numbers, and define

$$S = \sum_{k=1}^{\infty} a_k^2, \quad T = \sum_{k=1}^{\infty} k^2 a_k^2. \tag{1.4}$$

If either $S=\infty$ or $T=\infty$, there is nothing to prove. We may thus assume that both S and T are finite. If some a_k is non-zero (this is needed for strict inequality in the integral estimate below), we have by the Schwarz inequality

$$\begin{aligned}\left(\sum_{k=1}^{\infty} a_k\right)^2 &= \left(\sum_{k=1}^{\infty} a_k\sqrt{\alpha+\beta k^2}\frac{1}{\sqrt{\alpha+\beta k^2}}\right)^2 \\ &\leq \sum_{k=1}^{\infty} a_k^2(\alpha+\beta k^2)\sum_{k=1}^{\infty}\frac{1}{\alpha+\beta k^2} \\ &< (\alpha S+\beta T)\int_0^{\infty}\frac{1}{\alpha+\beta x^2}\,dx \\ &= \frac{\pi}{2}\left(S\sqrt{\frac{\alpha}{\beta}}+T\sqrt{\frac{\beta}{\alpha}}\right).\end{aligned}$$

Now, in view of (1.4), (1.1) follows if we just put $\alpha=T$ and $\beta=S$. □

Remark 1.6 Let

$$y=\frac{\alpha}{\beta}.$$

Then, according to the above proof,

$$\left(\sum_{k=1}^{\infty} a_k\right)^2 < C\left(y^{\frac{1}{2}}S+y^{-\frac{1}{2}}T\right). \tag{1.5}$$

In this *additive inequality*, y occurs in the two terms on the right-hand side, with powers *of opposite signs*. Thus, for fixed S and T, the right-hand side can be minimized with respect to y, and the minimum occurs at

$$y=\frac{T}{S}.$$

In this manner, we transform (1.5) to the *multiplicative inequality* (1.1). This simple but ingenious idea is the core of the above proof, and can be used to prove far more general versions of Carlson's inequality, as will be seen in later chapters of this book.

Also the second proof by Hardy is quite simple, although not as elementary as the first one, since it requires some knowledge about the theory of Fourier series.

Proof 2 (Hardy, 1936). Again, define S and T by (1.4), and let

$$f(x) = \sum_{k=1}^{\infty} a_k \cos kx.$$

If T converges, so does S, and by the Parseval identity it holds that

$$S = \frac{2}{\pi} \int_0^{\pi} f^2(x)\, dx \quad \text{and} \quad T = \frac{2}{\pi} \int_0^{\pi} f'^2(x)\, dx.$$

Moreover,

$$\int_0^{\pi} f(x)\, dx = 0, \tag{1.6}$$

so there exists a ξ between 0 and π for which $f(\xi) = 0$. Thus the Schwarz inequality yields

$$\begin{aligned}
\left(\sum_{k=1}^{\infty} a_k\right)^2 &= f^2(0) = f^2(0) - f^2(\xi) \\
&= 2 \int_{\xi}^{0} f(x) f'(x)\, dx \\
&\leq 2 \sqrt{\int_0^{\pi} f^2(x)\, dx} \sqrt{\int_0^{\pi} f'^2(x)\, dx} \\
&= 2 \sqrt{\frac{\pi}{2} S} \sqrt{\frac{\pi}{2} T} = \pi \sqrt{ST}.
\end{aligned}$$

Thus, in view of (1.4), the non-strict inequality (1.1) is proved. In order to prove the strict inequality, we first note that in the application of the Schwarz inequality above, equality holds if and only if $f'^2(x) = b^2 f^2(x)$ or

$$f(x) = ae^{bx}$$

for some constants a and b. In view of (1.6), this is possible only if $a = 0$, and in this case all the a_k must vanish. However, we have assumed that this is not the case, and hence the strict inequality (1.1) must hold. □

Remark 1.7 Also this idea of proof has been used to derive several generalizations and applications of (1.1). Here, we just mention the Ph. D. thesis of A. Kamaly [39] and refer to the references given there. (See also Section 5.3.6.)

1.3 An Alternate Proof

In Chapter 3, we will show that the integral version of Carlson's inequality implies the inequality (1.1). We also give some alternate proofs of the continuous version, thus also of (1.1). Here, however, we include a proof of the discrete version based on Hilbert's double sum inequality. Although this does not give the sharp constant π^2 in (1.1), it still gives the same qualitative information.

Proof by Hilbert's Inequality. The well-known Hilbert inequality states that

$$\sum_{k=1}^{\infty}\sum_{l=1}^{\infty}\frac{a_k b_l}{k+l} \leq \frac{\pi}{\sin\frac{\pi}{p}}\left(\sum_{k=1}^{\infty} a_k^p\right)^{\frac{1}{p}}\left(\sum_{l=1}^{\infty} b_l^{p'}\right)^{\frac{1}{p'}} \tag{1.7}$$

for all sequences $\{a_k\}_k$ and $\{b_l\}_l$ of non-negative numbers, with equality if and only if either all the a_k or all the b_l are zero. With $p = 2$ and $b_l = la_l$, it follows from (1.7) that

$$\begin{aligned}\left(\sum_{k=1}^{\infty} a_k\right)^2 &= \sum_{k=1}^{\infty} a_k \sum_{l=1}^{\infty} a_l \\ &= \sum_{k=1}^{\infty}\sum_{l=1}^{\infty}\frac{(k+l)a_k a_l}{k+l} \\ &= \sum_{k=1}^{\infty}\sum_{l=1}^{\infty}\frac{k a_k a_l}{k+l} + \sum_{k=1}^{\infty}\sum_{l=1}^{\infty}\frac{a_k l a_l}{k+l} \\ &< 2\pi\left(\sum_{k=1}^{\infty} a_k^2\right)^{\frac{1}{2}}\left(\sum_{l=1}^{\infty} l^2 a_l^2\right)^{\frac{1}{2}},\end{aligned}$$

provided not all a_k are zero. Squaring yields (1.1) with the constant π^2 replaced by $4\pi^2$. □

1.4 Carlson's Inequality for Finite Sums

Suppose that m is a positive integer. The inequality (1.1) holds, in particular, if $a_k = 0$ for $k > m$, i.e. if the series are replaced by finite sums with m terms. If, however, we restrict attention to finite sums with a fixed number of terms, the constant π^2 need no longer be (and is not) sharp. We have the following preliminary result by L. Larsson, Z. Páles and L.-E. Persson [52].

Proposition 1.1 Let $a_1, \dots, a_m$ be non-negative numbers, not all zero. Then

$$\left(\sum_{k=1}^{m} a_k\right)^4 < (2\arctan m)^2 \sum_{k=1}^{m} a_k^2 \sum_{k=1}^{m} k^2 a_k^2. \tag{1.8}$$

We see that the limiting case of (1.8) when $m \to \infty$ is (1.1).

Proof. Using the same notation and method as in Hardy's first proof above, we get

$$\left(\sum_{k=1}^{m} a_k\right)^2 < \int_0^m \frac{dx}{\alpha + \beta x} \cdot (\alpha S + \beta T).$$

The integral can be evaluated exactly:

$$\int_0^m \frac{dx}{\alpha + \beta x^2} = \frac{1}{\sqrt{\alpha\beta}} \arctan\left(\sqrt{\frac{\beta}{\alpha}} m\right).$$

Since $S \le T$, if we put $\alpha = T$ and $\beta = S$, we will always have

$$\arctan\left(\sqrt{\frac{\beta}{\alpha}} m\right) \le \arctan m,$$

so that

$$\begin{aligned}\left(\sum_{k=1}^{m} a_k\right)^2 &< \arctan m(\sqrt{ST} + \sqrt{ST}) \\ &= 2\arctan m \left(\sum_{k=1}^{m} a_k^2 \sum_{k=1}^{m} k^2 a_k^2\right)^{1/2}\end{aligned}$$

which, after squaring, yields the desired inequality. □

Remark 1.8 It should be mentioned that the constant $(2\arctan m)^2$ in (1.8), although strictly smaller than π^2 for each finite m, is not sharp. We consider the inequalities

$$\left(\sum_{k=1}^{m} a_k\right)^4 \le C_m \sum_{k=1}^{m} a_k^2 \sum_{k=1}^{m} k^2 a_k^2,$$

and seek, for each $m = 1, 2, \ldots$ the *sharp* constant C_m. Thus Proposition 1.1 says that $C_m \leq (2 \arctan m)^2$. It is clear, however, that

$$C_1 = 1 < \frac{\pi^2}{4} = (2 \arctan 1)^2.$$

Moreover, numerical calculations show that

$$\begin{aligned} C_2 &= \sup_{\alpha > 0} \frac{(1+\alpha)^4}{(1+\alpha^2)(1+4\alpha^2)} \\ &\approx 2.0311 < 4.9031 \approx (2 \arctan 2)^2. \end{aligned}$$

Note, also, that we can show that $C_2 \leq 4$, as follows. By convexity

$$\begin{aligned} (a+b)^4 &= [(a+b)^2]^2 \\ &\leq 4(a^2+b^2)^2 \\ &= 4(a^2+b^2)(a^2+b^2) \\ &\leq 4(a^2+b^2)(a^2+4b^2). \end{aligned}$$

This gives a slightly better result than Proposition 1.1 for the case $m = 2$. Furthermore, consider Carlson's observation regarding the application of the Hölder–Rogers inequality, but now to sums with m terms:

$$\left(\sum_{k=1}^{m} a_k \right)^4 \leq \left(\sum_{k=1}^{m} \frac{1}{k} \right)^2 \sum_{k=1}^{m} a_k^2 \sum_{k=1}^{m} k^2 a_k^2.$$

If $1 \leq m < 10$, it holds that

$$\sum_{k=1}^{m} \frac{1}{k} < 2 \arctan m,$$

so for such m, this method gives a better constant than Proposition 1.1.

Problem 1 Find a formula for

$$C_m, \quad m = 1, 2, \ldots.$$

Carlson type inequalities for finite sums will be discussed further in Section 2.10 of Chapter 2.

Chapter 2

Some Extensions and Complements of Carlson's Inequalities

In this chapter, we present some different variations of Carlson's inequalities, whose origins are spread in time from 1937 to 2005.

2.1 Gabriel

R. M. Gabriel [27] mentioned in a paper from 1937, that Hardy's method could be used to prove a more general version of Carlson's inequality. However, he chose to use a method similar to Carlson's original proof. We state Gabriel's result here.

Theorem 2.1 (Gabriel, 1937) If $p > 1$ and $0 < \delta \leq p - 1$, then

$$\left| \sum_{k=1}^{\infty} a_k \right|^{2p} < C \sum_{k=1}^{\infty} k^{p-1-\delta} |a_k|^p \sum_{k=1}^{\infty} k^{p-1+\delta} |a_k|^p, \tag{2.1}$$

where

$$C = \frac{4}{(2\delta)^{2p-2}} B\left(\frac{1}{2p-2}, \frac{1}{2p-2} \right)^{2p-2}. \tag{2.2}$$

Remark 2.1 With $p = 2$ and $\delta = 1$, (2.1) reduces to Carlson's inequality (1.1).

Remark 2.2 Note that this theorem allows the a_k to be complex. Although this is merely a notational matter, it is an interesting observation, since most authors on the area at this time understand the a_k to be real, sometimes without even mentioning this.

Remark 2.3 Gabriel also proved that the inequality (2.1) fails for any choice of C when $\delta = 0$. In fact, it suffices to consider the sequence $\{a_k\}_{k=1}^{\infty}$ defined by

$$a_k = \frac{1}{k}, \quad k = 1, \ldots, m,$$
$$a_k = 0, \quad k > m$$

and let $m \to \infty$.

2.2 Levin

V. I. Levin [55] gave another variation of Carlson's inequality (1.1). Instead of using two factors on the right-hand side of the inequality, he allowed any *odd* number of factors.

Theorem 2.2 (Levin, 1938) If $a_k \geq 0$, $k = 1, 2, \ldots$, and m is a positive integer, then

$$\left(\sum_{k=1}^{\infty} a_k\right)^{(m+1)(2m+1)} < C \prod_{j=0}^{2m} \sum_{k=1}^{\infty} k^j a_k^{m+1}, \tag{2.3}$$

unless all a_k are zero, where

$$C = \prod_{j=1}^{2m+1} j^{2j-2m-1} = (2m+1)^{2m+1} \prod_{j=0}^{2m} \binom{2m}{j}$$

is the sharp constant.

Remark 2.4 It is interesting to note that the sharp constant is an integer in every case covered by Theorem 2.2.

Example 2.1 In the cases $m = 1$ and $m = 2$, we get the following sharp inequalities.

$$\left(\sum_{k=1}^{\infty} a_k\right)^6 < 54 \sum_{k=1}^{\infty} a_k^2 \sum_{k=1}^{\infty} k a_k^2 \sum_{k=1}^{\infty} k^2 a_k^2,$$
$$\left(\sum_{k=1}^{\infty} a_k\right)^{15} < 300000 \sum_{k=1}^{\infty} a_k^3 \sum_{k=1}^{\infty} k a_k^3 \sum_{k=1}^{\infty} k^2 a_k^3 \sum_{k=1}^{\infty} k^3 a_k^3 \sum_{k=1}^{\infty} k^4 a_k^3.$$

Thus there is an inequality with the speed of light as sharp constant!

As a step in the proof, Levin used the following interesting observation.

Lemma 2.1 If m is a positive integer and $p_0 = 1$, $p_j > 0$, $j = 1, \ldots, 2m$, then

$$\int_0^\infty \left(\sum_{j=0}^{2m} \binom{2m}{j} p_j x^{2k-j} \right)^{-1/m} dx \le \left(\prod_{j=0}^{2k} p_j \right)^{-1/m(2m+1)}.$$

Proof of Theorem 2.2. If $c_0, \ldots, c_{2m}$ are any positive numbers, then, by writing

$$\begin{aligned} a_k = & \left(\sum_{j=0}^{2m} c_j k^j \right)^{-1/(m+1)} \\ & \times \left(\sum_{j=0}^{2m} c_j k^j \right)^{1/(m+1)} a_k, \end{aligned}$$

the Hölder–Rogers inequality with exponents

$$m+1 \quad \text{and} \quad \frac{m+1}{m}$$

implies that

$$\begin{aligned} \left(\sum_{k=1}^{\infty} a_k \right)^{(m+1)(2m+1)} < & \left(\int_0^\infty \frac{dx}{(c_0 + c_1 x + \ldots + c_{2m} x^{2m})^{1/m}} \right)^{m(2m+1)} \\ & \times \left(\sum_{j=0}^{2m} c_j S_j \right)^{m(2m+1)}, \end{aligned}$$

where the S_j denote the series on the right-hand side of (2.3). If the c_j are chosen so that

$$(c_j S_j)^{2m-1} = \prod_{i=0}^{2m} S_i,$$

this takes the form of (2.3), and Lemma 2.1 can be used to estimate the constant. By considering the sequence defined by

$$a_k = \frac{\gamma}{(1+\gamma k)^2}$$

and letting $\gamma \to 0$, the constant is shown to be sharp. □

2.3 Caton

W. B. Caton [24] generalized the idea of looking at Carlson's inequality (1.1) as a limiting case of the Hölder–Rogers inequality. He noted that if $p_i \in (0, \infty)$, $i = 1, 2, 3$, are such that

$$\sum_{i=1}^{3} \frac{1}{p_i} = 1,$$

if the non-negative numbers β_i, $i = 1, 2, 3$ satisfy

$$\beta_1 + \beta_2 = \beta_3,$$

and if

$$\alpha_1, \alpha_2 \geq 0, \quad \alpha_1 + \alpha_2 = 1,$$

then Hölder's inequality implies that

$$\sum_{k=1}^{\infty} a_k \leq C \left(\sum_{k=1}^{\infty} (k^{\beta_1} a_k^{\alpha_1})^{p_1} \right)^{1/p_1} \left(\sum_{k=1}^{\infty} (k^{\beta_2} a_k^{\alpha_2})^{p_2} \right)^{1/p_2}, \tag{2.4}$$

where

$$C = C(\beta_3, p_3) = \left(\sum_{k=1}^{\infty} k^{-\beta_3 p_3} \right)^{1/p_3}.$$

The constant C is finite when $\beta_3 p_3 > 1$, but tends to infinity as $\beta_3 p_3 \searrow 1$. Caton investigated under which conditions an inequality of the form (2.4), with a finite constant C, exists in the case $\beta_3 p_3 = 1$. More precisely, he gave a necessary condition for inequality, and a lower bound for the constant C in (2.4). The main result may be stated as follows.

Theorem 2.3 (Caton, 1940) Let $\sigma_i = \alpha_i p_i$ and $\tau_i = \beta_i p_i$, $i = 1, 2$. If (2.4) holds with $\sigma_1 = \sigma_2 = \sigma$, and $\tau_1 > \tau_2$, then necessarily

$$C \geq \frac{(\tau_1 - \tau_2)^{1-2/p_3}}{(\sigma - 1 - \tau_2)^{1/p_1} (\tau_1 + 1 - \sigma)^{1/p_2}}$$
$$\times B\left(\frac{1}{\tau_1 - \tau_2} - \frac{\tau_2}{(\sigma - 1)(\tau_1 - \tau_2)}, \frac{\tau_1}{(\sigma - 1)(\tau_1 - \tau_2)} - \frac{1}{\sigma - 1} \right)^{1/p_3},$$

where $B(\cdot, \cdot)$ denotes the Beta function.

2.4 Bellman

R. Bellman [9] extended Theorem 2.1 in 1943. He proved two versions of the inequality, collected in the following theorem.

Theorem 2.4 (Bellman, 1943)

(a) Suppose that $p, q > 1$, $\lambda, \mu > 0$. Then there is a constant C such that

$$\left(\sum_{k=1}^{\infty} a_k\right)^{p\mu+q\lambda} < C\left(\sum_{k=1}^{\infty} k^{p-1-\lambda} a_k^p\right)^{\mu}\left(\sum_{k=1}^{\infty} k^{q-1+\mu} a_l^q\right)^{\lambda}. \quad (2.5)$$

(b) If $\alpha, \beta > 1$, then there is a constant C such that

$$\left(\sum_{k=1}^{\infty} a_k\right)^{\alpha\beta+\alpha-\beta} < C\sum_{k=1}^{\infty} a_k^{\alpha}\left(\sum_{k=1}^{\infty} k^{\beta} a_k^{\beta}\right)^{\alpha-1}.$$

Remark 2.5 If we put $q = p$ and $\mu = \lambda$ in part (b) of Theorem 2.4, we get Theorem 2.1, and thus Theorem 1.1 as special cases. Theorem 1.1 also follows from part (a) if we put $\alpha = \beta = 2$.

Remark 2.6 Bellman also proved the corresponding continuous inequalities (see Theorem 3.2 of Chapter 3, where his proof of the first part is reconstructed). Although not stated explicitly in [9], the proof shows that in part (a) above, we may choose

$$C = \left(\frac{(p\mu+q\lambda)^{p\mu+q\lambda}}{(p\mu)^{p\mu}(q\lambda)^{q\lambda}}\right)^{1/pq} 2^{(pq-1)(p\mu+q\lambda)/pq}$$
$$B\left(\frac{\lambda}{p-1}, \frac{p-\lambda}{p-1}\right)^{\mu/(p-1)} B\left(\frac{\mu}{q-1}, \frac{q-\lambda}{q-1}\right)^{\lambda/(q-1)}.$$

Note that we need the additional assumptions $\lambda < p$ and $\mu < q$. Let us also mention that the sharp constant was later found by Levin [56] (see Chapter 4).

2.5 Two Discrete Carlson By-products

From Carlson's proof of Theorem 1.1, we extract the following sharpening of (1.1) (see Section 1.1 of Chapter 1 or inequality (B.8) of Appendix B).

Proposition 2.1 (Carlson, 1935) For any non-zero sequence $\{a_k\}_{k=1}^{\infty}$ of non-negative numbers, it holds that

$$\left(\sum_{k=1}^{\infty} a_k\right)^4 < \pi^2 \sum_{k=1}^{\infty} a_k^2 \sum_{k=1}^{\infty} \left(\left(k - \frac{1}{2}\right)^2 + \frac{3}{16}\right) a_k^2. \tag{2.6}$$

For completeness, we also extract the following double-series result from [23] (see inequality (B.10) of Appendix B).

Proposition 2.2 (Carlson, 1935) If $\{a_k\}_{k=1}^{\infty}$ and $\{b_k\}_{k=1}^{\infty}$ are non-zero sequences of non-negative numbers, then

$$\left(\sum_{k=1}^{\infty} a_k \sum_{k=1}^{\infty} b_k\right)^2 < \frac{\pi^2}{2} \left\{\sum_{k=1}^{\infty} a_k^2 \sum_{k=1}^{\infty} k^2 b_k^2 + \sum_{k=1}^{\infty} b_k^2 \sum_{k=1}^{\infty} k^2 a_k^2\right\}.$$

2.6 Landau and Levin–Stečkin

E. Landau (see the appendix [58] of the Russian translation of [34]; see also [33]) proved the following sharpening of (1.1):

$$\left(\sum_{k=1}^{\infty} a_k\right)^4 < \pi^2 \sum_{k=1}^{\infty} a_k^2 \sum_{k=1}^{\infty} \left(k - \frac{1}{2}\right)^2 a_k^2. \tag{2.7}$$

In fact, it is slightly stronger than the inequality (2.6). Moreover, in the appendix mentioned above, the following result was proved by V. I. Levin and S. B. Stečkin [58]:

Theorem 2.5 (Levin–Stečkin, 1960) If

$$p > \frac{1}{8} \quad \text{and} \quad 0 < q^2 \leq \frac{8p^2 - p}{4(p+1)}, \tag{2.8}$$

then the inequality

$$\left(\sum_{k=1}^{\infty} a_k\right)^{2p+2} < C \sum_{k=1}^{\infty} \left(k - \frac{1}{2}\right)^{p-q} a_k^{p+1} \sum_{k=1}^{\infty} \left(k - \frac{1}{2}\right)^{p+q} a_k^{p+1} \tag{2.9}$$

of Landau type holds. Here, we may choose

$$C = C(p,q) = 4(2q)^{-2p} B\left(\frac{1}{2p}, \frac{1}{2p}\right)^{2p}, \tag{2.10}$$

and this constant is sharp.

Here, as usual, $B(\cdot,\cdot)$ denotes the Beta-function.

Remark 2.7 Note that in order to get (2.7) from (2.9), we would need to put $p = q = 1$, but this case is *not* covered by (2.8).

In the next section, an extension of the above result is given. However, we sketch a proof of Theorem 2.5 here.

Let

$$S_0 = \sum_{k=1}^{\infty} \left(k - \frac{1}{2}\right)^{p-q} a_k^{p+1}$$

and

$$S_1 = \sum_{k=1}^{\infty} \left(k - \frac{1}{2}\right)^{p+q} a_k^{p+1}.$$

We note first that if the conditions (2.8) hold, then the function

$$f(y) = \left(\lambda y^{p-q} + \frac{1}{\lambda} y^{p+q}\right)^{-1/p}, \quad y > 0$$

is convex for any $\lambda > 0$. Thus, for any positive integer k, it holds by Hadamard's inequality that

$$f\left(k - \frac{1}{2}\right) < \int_{k-1}^{k} f(y)\, dy.$$

Moreover, we have

$$\int_0^{\infty} \left(\lambda y^{p-q} + \frac{1}{\lambda} y^{p+q}\right)^{-1/p} dy = \frac{1}{2q} B\left(\frac{1}{2p}, \frac{1}{2p}\right).$$

Thus, if we write

$$a_k = \left(\lambda\left(k - \frac{1}{2}\right)^{p-q} + \frac{1}{\lambda}\left(k - \frac{1}{2}\right)^{p+q}\right)^{-1/(p+1)}$$
$$\left(\lambda\left(k - \frac{1}{2}\right)^{p-q} + \frac{1}{\lambda}\left(k - \frac{1}{2}\right)^{p+q}\right)^{1/(p+1)} a_k,$$

it follows by the Hölder–Rogers inequality with exponents

$$\frac{p+1}{p} \quad \text{and} \quad p+1$$

that

$$\left(\sum_{k=1}^{\infty} a_k\right)^{2p+2} \leq S\left(\lambda S_0 + \frac{1}{\lambda} S_1\right)^2,$$

where

$$\begin{aligned} S &= \left[\sum_{k=1}^{\infty} \left(\lambda\left(k-\frac{1}{2}\right)^{p-q} + \frac{1}{\lambda}\left(k-\frac{1}{2}\right)^{p+q}\right)^{-1/p}\right]^{2p} \\ &= \left(\sum_{k=1}^{\infty} f\left(k-\frac{1}{2}\right)\right)^{2p} \\ &< \left(\sum_{k=1}^{\infty} \int_{k-1}^{k} f(y)\,dy\right)^{2p} \\ &= \left(\int_0^{\infty} f(y)\,dy\right)^{2p} \\ &= \left(\frac{1}{2q} B\left(\frac{1}{2p}, \frac{1}{2p}\right)\right)^{2p}. \end{aligned}$$

Letting

$$\lambda = \sqrt{\frac{S_1}{S_0}},$$

we get (2.9) with C given by (2.10).

2.7 Some Extensions of the Landau and Levin–Stečkin Inequalities

In this section, Theorem 2.5 is extended to allow more flexibility of parameters. One may ask, for instance, whether there is a constant C such that the inequality

$$\left(\sum_{k=1}^{\infty} a_k\right)^{5/2} < C \sum_{k=1}^{\infty} a_k^{5/4} \sum_{k=1}^{\infty} \left(k-\frac{1}{2}\right)^{1/2} a_k^{5/4} \tag{2.11}$$

holds for all non-zero sequences $\{a_k\}_{k=1}^{\infty}$ of non-negative numbers. Theorem 2.6 will give an affirmative answer to this. In addition, it is pointed out that the constant $C = \frac{4}{\sqrt{3}}$ is sharp. This inequality can be used to establish convergence of $\sum a_k$ in situations where (2.7) fails. (Choose e.g. $a_k = (k-1/2)^{-7/5}$.)

For a fixed sequence $a_1, a_2, \ldots$ of non-negative numbers, and any real numbers σ and τ, we define

$$S(\sigma,\tau) = \sum_{k=1}^{\infty}\left(k-\frac{1}{2}\right)^{\sigma} a_k^{\tau}. \tag{2.12}$$

Thus, for example, the inequality (2.9) of Theorem 2.5 can be written

$$S(0,1)^{2p+2} < C \cdot S(p-q,p+1)S(p+q,p+1).$$

In what follows, we present a generalization and unification of the inequality (2.7) and Theorem 2.5 by L. Larsson, J. Pečarić and L.-E. Persson [53].

2.7.1 *The Case* $p = 1$

As pointed out in Remark 2.7, Theorem 2.5 does not cover (2.7). For illustrative purposes, we present the following preliminary result, which fills the gap between the inequality (2.7) and Theorem 2.5.

Proposition 2.3 Let $q > 0$. Then there exists a constant C such that

$$S(0,1)^4 < C \cdot S(1-q,2)S(1+q,2).$$

We may choose

$$C = \begin{cases} \frac{\pi^2}{q^2}, & 0 < q \le \sqrt{\frac{7}{8}}, \\ \frac{8}{7}\pi^2, & \sqrt{\frac{7}{8}} < q < 1, \\ \pi^2, & q \ge 1. \end{cases}$$

The constant is sharp in the cases $0 < q \le \sqrt{\frac{7}{8}}$ and $q = 1$.

The proof, which is postponed to the next section, uses inequality (2.7) for the case $q = 1$ and Theorem 2.5 for the case $0 < q \le \sqrt{\frac{7}{8}}$. For the remaining cases, an interpolation argument is used.

Remark 2.8 The sharp constant in the above result is unknown in the case $\sqrt{\frac{7}{8}} < q < 1$. By investigating the form of the sharp constant in known cases, one is tempted to *guess* that

$$C = \frac{\pi^2}{q^2}$$

is the sharp constant whenever $0 < q \leq 1$. However, this remains an open question.

Problem 2 Find the sharp constant in Proposition 2.3 for

$$\sqrt{\frac{7}{8}} < q \neq 1.$$

2.7.2 *General p*

The following result extends Theorem 2.5 in the sense that we introduce a new parameter r. More precisely, we consider the inequality

$$S(0,1)^{2r+2} < C \cdot S(p-q, r+1)^{(q-(r-p))/q} S(p+q, r+1)^{(q+(r-p))/q}. \quad (2.13)$$

The result is the following.

Theorem 2.6 Suppose that $r \geq p$. Then a necessary and sufficient condition for the existence of a finite constant C such that the inequality (2.13) holds is that

$$q > r - p. \quad (2.14)$$

If

$$r - p < q \leq q_0,$$

where $q_0 > 0$ is defined by

$$q_0^2 = \max\left\{\frac{p(p+r)}{2r+1}, \frac{4p(p+r)-r}{4(r+1)}\right\},$$

then we may choose

$$C = C(p,q,r) = 4\left(\frac{1}{2q}B\left(\frac{q-(r-p)}{2rq}, \frac{q+(r-p)}{2rq}\right)\right)^{2r},$$

and this constant is sharp. In the case

$$q > q_0,$$

we may choose

$$C = C(p, q_0, r).$$

Remark 2.9 If we put $r = p$ in Theorem 2.6, we get an extended version of Theorem 2.5. If, in addition,

$$p < \frac{1}{2},$$

then there are cases which give the best constant but which are not covered by earlier results. Indeed, let

$$r = p = q$$

in Theorem 2.6. We then get the inequality

$$\left(\sum_{k=1}^{\infty} a_k\right)^{2q+2} < 4\left(\frac{1}{2q}B\left(\frac{1}{2q}, \frac{1}{2q}\right)\right)^{2q} \sum_{k=1}^{\infty} a_k^{q+1} \sum_{k=1}^{\infty}\left(k - \frac{1}{2}\right)^{2q} a_k^{q+1}$$

for all non-zero sequences of non-negative numbers, and the constant is sharp. In particular, when $q = \frac{1}{4}$, we obtain the inequality (2.11) with the sharp constant

$$C = \frac{4}{\sqrt{3}},$$

as announced.

Remark 2.10 Let us mention that the sharp constant in Theorem 2.6 is unknown in the case $q > q_0$.

Problem 3 Find the sharp constant in Theorem 2.6 for $q > q_0$.

2.8 Proofs

For the proof of Theorem 2.6, we need two lemmas. We apply a convexity argument similar to that used for Theorem 2.5. This is stated in the first lemma below. Its proof is elementary, and is not included here. See, however, Lemma 2 in [53].

Lemma 2.2 Let

$$f(y) = \frac{1}{(y^{p-q} + y^{p+q})^{1/r}}, \quad y > 0.$$

If p, q and r are such that

$$q^2 \leq \max\left\{\frac{p(p+r)}{2r+1}, \frac{4p(p+r)-r}{4(r+1)}\right\}, \tag{2.15}$$

then the function f is convex.

Our second lemma is another type of convexity result, in the guise of an interpolation inequality. This is what we need to prove Propostion 2.3. This was inspired by *the Kjellberg Principle* (see Proposition 3.2 of Chapter 3).

Lemma 2.3 Let $S(\sigma,\tau)$ be as defined by (2.12). If $0<\theta<1$ and σ_i,τ_i, $i=0,1$ are real numbers, then

$$S\big((1-\theta)\sigma_0+\theta\sigma_1,(1-\theta)\tau_0+\theta\tau_1\big) \leq S(\sigma_0,\tau_0)^{1-\theta}S(\sigma_1,\tau_1)^{\theta}.$$

Proof. The Hölder–Rogers inequality with exponents

$$\frac{1}{1-\theta} \quad \text{and} \quad \frac{1}{\theta}$$

yields the desired inequality. □

We can now prove our extensions of the Landau and Levin–Stečkin results.

Proof of Proposition 2.3. The case $0<q^2\leq\frac{7}{8}$ follows from Theorem 2.5 if we put $p=1$. Suppose that

$$\frac{7}{8}<q^2<1,$$

and let

$$\eta=\frac{1}{2}-\frac{1}{2q}\sqrt{\frac{7}{8}} \quad \text{and} \quad \theta=\frac{1}{2}+\frac{1}{2q}\sqrt{\frac{7}{8}}.$$

Then, since

$$1+\sqrt{\frac{7}{8}}=(1-\theta)(1-q)+\theta(1+q)$$

and

$$1-\sqrt{\frac{7}{8}}=(1-\eta)(1-q)+\eta(1+q),$$

it follows from Lemma 2.3 and the case $q = \sqrt{\frac{7}{8}}$, that

$$\begin{aligned} S(0,1)^4 &< C\cdot S\left(1-\sqrt{\frac{7}{8}},2\right)S\left(1+\sqrt{\frac{7}{8}},2\right) \\ &\leq C\cdot S(1-q,2)^{1-\eta}S(1+q,2)^{\eta} \\ &\qquad S(1-q,2)^{1-\theta}S(1+q,2)^{\theta} \\ &= C\cdot S(1-q,2)S(1+q,2), \end{aligned}$$

where

$$C = C\left(1, \sqrt{\frac{7}{8}}\right) = \frac{8}{7}\pi^2.$$

The case $q = 1$ is precisely (2.7). Finally, for the case $q > 1$, let

$$\theta = \frac{q+1}{2q} \quad \text{and} \quad \eta = \frac{q-1}{2q}.$$

Then

$$0 = (1-\eta)(1-q) + \eta(1+q) \quad \text{and} \quad 2 = (1-\theta)(1-q) + \theta(1+q),$$

so again by Lemma 2.3,

$$\begin{aligned} S(0,1)^4 &< \pi^2 S(0,2)S(2,2) \\ &\leq \pi^2[S(1-q,2)^{1-\eta}S(1+q,2)^{\eta}S(1-q,2)^{1-\theta}S(1+q,2)^{\theta}] \\ &= \pi^2 S(1-q,2)S(1+q,2). \end{aligned}$$

The proof is complete. □

Proof of Theorem 2.6. Suppose first that

$$r - p < q \leq q_0,$$

and let $\lambda > 0$. It is clear that if g is defined by

$$\begin{aligned} g(y) &= \lambda^{-\frac{p}{rq}} f(\lambda^{-1/q}y) \\ &= \left(\lambda y^{p-q} + \frac{1}{\lambda}y^{p+q}\right)^{-\frac{1}{r}}, \quad y > 0, \end{aligned}$$

where f is as in Lemma 2.2, then g is convex whenever f is. As in the proof of Theorem 2.5, for any values of the parameters p, q, and r for which g is convex, we can apply Hadamard's inequality so that for any k it holds that

$$g\left(k-\frac{1}{2}\right) < \int_{k-1}^{k} g(y)\,dy.$$

It follows, then, from the Hölder–Rogers inequality with exponents

$$\frac{r+1}{r} \quad \text{and} \quad r+1,$$

that

$$\begin{aligned}
S(0,1) &= \sum_{k=1}^{\infty} a_k \\
&= \sum_{k=1}^{\infty} \left(\lambda\left(k-\frac{1}{2}\right)^{p-q} + \frac{1}{\lambda}\left(k-\frac{1}{2}\right)^{p+q}\right)^{-\frac{1}{r+1}} \\
&\qquad \cdot \left(\lambda\left(k-\frac{1}{2}\right)^{p-q} + \frac{1}{\lambda}\left(k-\frac{1}{2}\right)^{p+q}\right)^{\frac{1}{r+1}} a_k \\
&\leq \left[\sum_{k=1}^{\infty} \left(\lambda\left(k-\frac{1}{2}\right)^{p-q} + \frac{1}{\lambda}\left(k-\frac{1}{2}\right)^{p+q}\right)^{-\frac{1}{r}}\right]^{\frac{r}{r+1}} \\
&\qquad \left[\sum_{k=1}^{\infty} \left(\lambda\left(k-\frac{1}{2}\right)^{p-q} + \frac{1}{\lambda}\left(k-\frac{1}{2}\right)^{p+q}\right) a_k^{r+1}\right]^{\frac{1}{r+1}} \\
&= \left[\sum_{k=1}^{\infty} g\left(k-\frac{1}{2}\right)\right]^{\frac{r}{r+1}} \left[\lambda S(p-q,r+1) + \frac{1}{\lambda} S(p+q,r+1)\right]^{\frac{1}{r+1}} \\
&< \left[\int_0^{\infty} g(y)\,dy\right]^{\frac{r}{r+1}} \left[\lambda S(p-q,r+1) + \frac{1}{\lambda} S(p+q,r+1)\right]^{\frac{1}{r+1}}.
\end{aligned}$$

In the integral above, we can subsitute $\lambda^{1/q} t^{1/2q}$ for y, then put $t = x/(1-x)$, which yields

$$\begin{aligned}
\int_0^{\infty} \left(\lambda y^{p-q} + \frac{1}{\lambda} y^{p+q}\right)^{-\frac{1}{r}} dy &= \frac{\lambda^{\frac{r-p}{rq}}}{2q} \int_0^{\infty} t^{\frac{q+(r-p)}{2rq}} (1+t)^{-\frac{1}{r}} \frac{dt}{t} \\
&= \frac{\lambda^{\frac{r-p}{rq}}}{2q} \int_0^{1} (1-x)^{\frac{q-(r-p)}{2rq}} x^{\frac{q+(r-p)}{2rq}} \frac{dx}{(1-x)x} \\
&= \frac{\lambda^{\frac{r-p}{rq}}}{2q} B\left(\frac{q-(r-p)}{2rq}, \frac{q+(r-p)}{2rq}\right).
\end{aligned}$$

Note that the integral converges, since by assumption $p-q<r<p+q$. Letting

$$\lambda=\sqrt{\frac{S(p+q,r+1)}{S(p-q,r+1)}},$$

we get (2.13) with

$$C=4\left(\frac{1}{2q}B\left(\frac{q-(r-p)}{2rq},\frac{q+(r-p)}{2rq}\right)\right)^{2r},$$

as desired. Furthermore, by comparing the cases of equality in the application of the Hölder–Rogers inequality above, we find that this constant is sharp.

Suppose next that $q>q_0$, and define

$$\theta=\frac{q+q_0}{2q}\quad\text{and}\quad\eta=\frac{q-q_0}{2q}.$$

Thus

$$p+q_0=(1-\theta)(p-q)+\theta(p+q)$$

and

$$p-q_0=(1-\eta)(p-q)+\eta(p+q),$$

and hence by Lemma 2.3, together with what has been proved above, it follows that

$$\begin{aligned}S(0,1)^{2r+2}&<C(p,q_0,r)S(p-q_0,r+1)^{\frac{q_0-(r-p)}{q_0}}S(p+q_0,r+1)^{\frac{q_0+(r-p)}{q_0}}\\&\leq C(p,q_0,r)[S(p-q,r+1)^{1-\eta}S(p+q,r+1)^{\eta}]^{\frac{q_0-(r-p)}{q_0}}\\&\qquad[S(p-q,r+1)^{1-\theta}S(p+q,r+1)^{\theta}]^{\frac{q_0+(r-p)}{q_0}}\\&=C(p,q_0,r)S(p-q,r+1)^{\frac{q-(r-p)}{q}}S(p+q,r+1)^{\frac{q+(r-p)}{q}},\end{aligned}$$

which is (2.13) with

$$C=C(p,q_0,r).$$

Thus the condition (2.14) is sufficient for the inequality (2.13) to hold for some constant C.

Suppose now that (2.13) holds for some constant C, and assume that $q = r - p$. Then the inequality under consideration reads

$$\left(\sum_{k=1}^{\infty} a_k\right)^{p+q+1} < C^{1/2} \sum_{k=1}^{\infty} \left(k - \frac{1}{2}\right)^{p+q} a_k^{p+q+1}. \tag{2.16}$$

Define

$$a_k = 0, \quad k = 1, 2$$

and

$$a_k = \left(\left(k - \frac{1}{2}\right) \log\left(k - \frac{1}{2}\right)\right)^{-1}, \quad k \geq 3.$$

Then

$$\sum_{k=1}^{\infty} a_k = \infty,$$

while

$$\sum_{k=1}^{\infty} \left(k - \frac{1}{2}\right)^{p+q} a_k^{p+q+1} = \sum_{k=3}^{\infty} \left(k - \frac{1}{2}\right)^{-1} \left(\log\left(k - \frac{1}{2}\right)\right)^{-p-q-1} < \infty.$$

It follows that (2.16) does not hold, which means that (2.13) does not hold in general. Moreover, it cannot hold for $q < r - p$ either, since then by what has been proved above together with Lemma 2.3, it would hold for $q = r - p$ as well. The proof is complete. □

2.9 Levin–Godunova

In 1965, Levin extended his Theorem 2.2 in cooperation with E. K. Godunova [57]. By basically the same method as Levin used for Theorem 2.2, they were able to overcome the restriction of using an odd number of factors on the right-hand side. Also, their inequality allows more flexibility in parameters, although the sharp constant is not found in the most general case. We recover the results below. In the proofs of the two main results, two different lemmas are used. However, the methods are similar, and we therefore give the details for the first theorem only. Throughout this section, m is some fixed positive integer.

Lemma 2.4 Let $c_j > 0$ and $\lambda_j > 0$, $j = 1, \ldots, m$, and put

$$p = \frac{1}{m}\sum_{j=1}^{m}\lambda_j,$$

$$r_1 = \frac{1}{s}\sum_{j=1}^{s}\lambda_j, \quad \text{and}$$

$$r_2 = \frac{1}{m-s}\sum_{j=s+1}^{m}\lambda_j,$$

where s is some integer between 0 and m. Suppose, moreover, that the λ_j are such that $r_2 > r_1$. Then

$$I = \int_0^\infty \frac{dx}{(c_1x^{\lambda_1} + \ldots + c_mx^{\lambda_m})^{1/p}} \le A\left(\prod_{j=1}^{m} c_j\right)^{-1/mp},$$

where

$$A = \min_{1\le s\le m-1} \frac{1}{r_2 - r_1}\frac{1}{[s^s(m-s)^{m-s}]^{1/mp}}B\left(\frac{m-s}{mp}, \frac{s}{mp}\right). \tag{2.17}$$

Remark 2.11 In the above lemma, we need to assume that the λ_j are so that $r_2 > r_1$ for *any* choice of s. This can be achieved, for instance, by arranging the numbers in increasing order: $\lambda_1 < \cdots < \lambda_m$. This requires the λ_j to be distinct. This is not a major restriction, however, since if two of the λ_j are equal, we may replace them by a single λ by adding the corresponding c_j.

Proof of Lemma 2.4. Define D_1 and D_2 by

$$D_1 = s\left(\prod_{j=1}^{s} c_j\right)^{1/s} \quad \text{and} \quad D_2 = (m-s)\left(\prod_{j=s+1}^{m} c_j\right)^{1/(m-s)}.$$

Thus, by the arithmetic–geometric mean inequality

$$\begin{aligned} c_1x^{\lambda_1} + \ldots + c_sx^{\lambda_s} &= s\frac{c_1x^{\lambda_1} + \ldots + c_sx^{\lambda_s}}{s} \\ &\ge s\left(c_1x^{\lambda_1}\cdots c_sx^{\lambda_s}\right)^{1/s} \\ &= s(c_1\cdots c_s)^{1/s}x^{(\lambda_1+\ldots+\lambda_s)/s} = D_1x^{r_1}, \end{aligned}$$

and similarly

$$c_{s+1}x^{\lambda_{s+1}} + \ldots + c_m x^{\lambda_m} \geq D_2 x^{r_2},$$

so that

$$I \leq \int_0^\infty \frac{dx}{(D_1 x^{r_1} + D_2 x^{r_2})^{1/p}}.$$

This integral can be evaluated to

$$\frac{\left(D_1^{p-r_2} D_2^{r_1-p}\right)^{1/p(r_2-r_1)}}{r_2 - r_1} B\left(\frac{p - r_1}{p(r_2 - r_1)}, \frac{r_2 - p}{p(r_2 - r_1)}\right).$$

Note that

$$\begin{aligned}
p - r_2 &= \frac{1}{m}\sum_{j=1}^{m} \lambda_j - \frac{1}{m-s}\sum_{j=s+1}^{m} \lambda_j \\
&= \frac{1}{m}\sum_{j=1}^{s} \lambda_j - \left(\frac{1}{m-s} - \frac{1}{m}\right)\sum_{j=s+1}^{m} \lambda_j \\
&= \frac{s}{m}\frac{1}{s}\sum_{j=1}^{s} \lambda_j - \frac{s}{m}\frac{1}{m-s}\sum_{j=s+1}^{m} \lambda_j \\
&= \frac{s}{m}(r_1 - r_2),
\end{aligned}$$

or

$$\frac{p - r_2}{r_2 - r_1} = -\frac{s}{m},$$

and in the same way we get

$$\frac{r_1 - p}{r_2 - r_1} = -\frac{m - s}{m}.$$

It follows that

$$\begin{aligned}
I \leq& \frac{1}{r_2 - r_1} s^{-s/mp}(c_1 \cdots c_s)^{-1/mp} \\
&\times (m - s)^{-(m-s)/mp}(c_{s+1} \cdots c_m)^{-1/mp} B\left(\frac{m - s}{mp}, \frac{s}{mp}\right) \\
=& \frac{1}{r_2 - r_1} \frac{1}{[s^s(m - s)^{m-s}]^{1/mp}} B\left(\frac{m - s}{mp}, \frac{s}{mp}\right) \left(\prod_{j=1}^{m} c_j\right)^{-1/mp}.
\end{aligned}$$

Since this can be done for any s, the assertion follows. □

The first theorem from [57] goes as follows.

Theorem 2.7 (Levin–Godunova, 1965) Suppose that $\lambda_j > 0$, $j = 1, \dots, m$, and define

$$p = \frac{1}{m} \sum_{j=1}^{m} \lambda_j.$$

Then for any non-zero sequence $a_1, a_2, \dots$ of non-negative numbers, the inequality

$$\left(\sum_{k=1}^{\infty} a_k \right)^{(p+1)m} < C \prod_{j=1}^{m} \sum_{k=1}^{\infty} k^{\lambda_j} a_k^{p+1} \tag{2.18}$$

holds, where we can choose

$$C = A^{pm} m^m,$$

where A is defined by (2.17).

Remark 2.12 If $m = 2M+1$ is odd and if $\lambda_j = j-1$ in the above theorem, then

$$p = \frac{1}{2M+1} \sum_{j=1}^{2M+1} (j-1) = M.$$

This reduces (2.18) to the inequality (2.3) of Theorem 2.2. However, Theorem 2.7 might not give the sharp constant.

Remark 2.13 As opposed to Theorem 2.2, Theorem 2.7 implies Carlson's inequality (1.1). Indeed, if $m = 2$, $\lambda_1 = p - q$, and $\lambda_2 = p + q$, then Theorem 2.7 reduces to Theorem 2.1 (with sharp constant). In particular, we get (1.1) if we put $m = 2$, $\lambda_1 = 0$, and $\lambda_2 = 2$.

Proof of Theorem 2.7. For $j = 1, \dots, m$, let

$$S_j = \sum_{k=1}^{\infty} k^{\lambda_j} a_k^{p+1},$$

and put

$$c_j = S_j^{-1} \left(\prod_{i=1}^{m} S_i \right)^{1/m}.$$

Thus

$$\sum_{j=1}^{m} c_j S_j = m \left(\prod_{j=1}^{m} S_j \right)^{1/m}.$$

The Hölder–Rogers inequality implies, then, that

$$\begin{aligned}\left(\sum_{k=1}^{\infty} a_k\right)^{(p+1)m} &\leq \left(\sum_{k=1}^{\infty}(c_1 k^{\lambda_1} + \ldots + c_m k^{\lambda_m})^{-1/p}\right)^{pm} \\ &\times \left(\sum_{k=1}^{\infty}(c_1 k^{\lambda_1} + \ldots + c_m k^{\lambda_m}) a_k^{p+1}\right)^{m} \\ &< \left(\int_0^{\infty} \frac{dx}{(c_1 x^{\lambda_1} + \ldots + c_m x^{\lambda_m})^{1/p}}\right)^{pm} m^m \prod_{j=1}^{m} S_j.\end{aligned}$$

We apply Lemma 2.4 and note that the choice of c_j implies that

$$\prod_{j=1}^{m} c_j = 1,$$

whereupon the desired result follows. □

The second result is yet another variation on Carlson's inequality. The proof is almost identical to that of Theorem 2.7, and also of Theorem 2.2, and we therefore omit it. However, we state the corresponding lemma needed for the proof.

Lemma 2.5 Suppose that $p_0 = 1$ and $p_j > 0$, $j = 1, \ldots, m$. Put

$$\lambda_j = \frac{m}{2} + h\left(j - \frac{m}{2}\right),$$

where $h > 0$. Then

$$\int_0^{\infty} \left(\sum_{j=0}^{m} \binom{m}{j} p_j x^{\lambda_j} \right)^{-2/m} dx \leq \frac{1}{h} \left(\prod_{j=1}^{m} p_j \right)^{-2/m(m+1)},$$

with equality if and only if for some p, $p_j = p^j$, $j = 1, \ldots, m$.

Theorem 2.8 (Levin–Godunova, 1965) Let λ_j be as in the above lemma. Then

$$\left(\sum_{k=1}^{\infty} a_k\right)^{(k+1)(k+2)/2} < C \prod_{j=0}^{m} \sum_{k=1}^{\infty} k^{\lambda_j} a_k^{(m+2)/2},$$

where

$$C = h^{-m(m+1)/2}(m+1)^{m+1} \prod_{j=0}^{m} \binom{m}{j}$$

is the sharp constant.

2.10 More About Finite Sums

The following two results by L. Larsson, Z. Páles and L.-E. Persson [52] concern finite sums rather than infinite series, and extend the discussion in Section 1.4 of Chapter 2.

For $\alpha > 0$, $\beta > 0$ and $0 < t < 1$, we define *the truncated Beta function* by

$$B(\alpha, \beta; t) = \int_0^t (1-s)^{\alpha-1} s^{\beta-1}\, ds = \int_0^t (1-s)^{\alpha} s^{\beta} \frac{ds}{(1-s)s}.$$

Suppose that m is a positive integer. Then the inequality in (a) of Theorem 2.4 holds, with the same constant (or even with Levin's *sharp* constant — see (4.9) in Chapter 4), for sequences $\{a_k\}_{k=1}^{\infty}$ for which $a_k = 0$ whenever $k > m$. However, if we consider fixed-length sequences only, the constant can be taken strictly smaller.

Theorem 2.9 Suppose that m is a positive integer, and that $r > 0$, and $\alpha_2 > \alpha_1 \geq 0$. If $a_1, \ldots, a_m$ are non-negative numbers, not all zero, then

$$\sum_{k=1}^{m} a_k < C \left(\sum_{k=1}^{m} k^{\alpha_1} a_k^{r+1}\right)^{\frac{\alpha_2 - r}{(\alpha_2-\alpha_1)(r+1)}} \left(\sum_{k=1}^{m} k^{\alpha_2} a_k^{r+1}\right)^{\frac{-\alpha_1 + r}{(\alpha_2-\alpha_1)(r+1)}}, \quad (2.19)$$

where we may choose

$$C = 2^{\frac{1}{r+1}} \left[\frac{1}{\alpha_2 - \alpha_1} B\left(\frac{\alpha_2 - r}{(\alpha_2-\alpha_1)r}, \frac{-\alpha_1 + r}{(\alpha_2-\alpha_1)r}; \frac{m^{\alpha_2-\alpha_1}}{1+m^{\alpha_2-\alpha_1}}\right)\right]^{\frac{r}{r+1}}. \quad (2.20)$$

Remark 2.14 The constant C defined by (2.20), which depends on m, tends to the sharp constant in (2.5) when $m \to \infty$. In the special case $r = 1$, $\alpha_1 = 0$, $\alpha_2 = 2$, we have

$$B\left(\frac{1}{2}, \frac{1}{2}; \frac{m^2}{1+m^2}\right) = 2 \arcsin \frac{m}{\sqrt{1+m^2}} = 2 \arctan m,$$

so that the inequality (1.8) follows as a special case of Theorem 2.9. When $m \to \infty$, this constant tends to the sharp constant $\sqrt{\pi}$ in (1.1). This special case is just Proposition 1.1.

Proof of Theorem 2.9. Let

$$S = \sum_{k=1}^{m} k^{\alpha_1} a_k^{r+1} \quad \text{and} \quad T = \sum_{k=1}^{m} k^{\alpha_2} a_k^{r+1},$$

and let λ be any positive number. If we write

$$a_k = (\lambda k^{\alpha_1} + \frac{1}{\lambda} k^{\alpha_2})^{-\frac{1}{r+1}} \cdot (\lambda k^{\alpha_1} + \frac{1}{\lambda} k^{\alpha_2})^{\frac{1}{r+1}} a_k,$$

we get by the Hölder–Rogers inequality

$$\begin{aligned}\left(\sum_{k=1}^{m} a_k\right)^{r+1} &\leq \left(\sum_{k=1}^{m} (\lambda k^{\alpha_1} + \lambda^{-1} k^{\alpha_2})^{-\frac{1}{r}}\right)^{r} \left(\sum_{k=1}^{m} (\lambda k^{\alpha_1} + \lambda^{-1} k^{\alpha_2}) a_k^{r+1}\right) \\ &= \left(\sum_{k=1}^{m} (\lambda k^{\alpha_1} + \lambda^{-1} k^{\alpha_2})^{-\frac{1}{r}}\right)^{r} (\lambda S + \lambda^{-1} T).\end{aligned} \tag{2.21}$$

Since the function

$$x \mapsto (\lambda x^{\alpha_1} + \lambda^{-1} x^{\alpha_2})^{-\frac{1}{r}}$$

is non-increasing for positive x, the last sum in (2.21) can be estimated by an integral. In fact, by making the substitution

$$x = \lambda^{\frac{2}{\alpha_2 - \alpha_1}} \left(\frac{s}{1-s}\right)^{\frac{1}{\alpha_2 - \alpha_1}},$$

we find that

$$
\begin{aligned}
\sum_{k=1}^{m} (\lambda k^{\alpha_1} + \lambda^{-1} k^{\alpha_2})^{-\frac{1}{r}} &< \int_0^m \frac{dx}{(\lambda x^{\alpha_1} + \lambda^{-1} x^{\alpha_2})^{\frac{1}{r}}} \\
&= \frac{\lambda^{\frac{2r-(\alpha_1+\alpha_2)}{(\alpha_2-\alpha_1)r}}}{\alpha_2 - \alpha_1} \int_0^{\frac{m^{\alpha_2-\alpha_1}}{\lambda^2+m^{\alpha_2-\alpha_1}}} (1-s)^{\frac{\alpha_2-r}{(\alpha_2-\alpha_1)r}} s^{\frac{-\alpha_1+r}{(\alpha_2-\alpha_1)r}} \frac{ds}{(1-s)s} \\
&= \frac{\lambda^{\frac{r-\frac{\alpha_1+\alpha_2}{2}}{\frac{\alpha_2-\alpha_1}{2}r}}}{\alpha_2-\alpha_1} B\left(\frac{\alpha_2 - r}{(\alpha_2-\alpha_1)r}, \frac{-\alpha_1+r}{(\alpha_2-\alpha_1)r}; \frac{m^{\alpha_2-\alpha_1}}{\lambda^2 + m^{\alpha_2-\alpha_1}} \right).
\end{aligned} \tag{2.22}
$$

The function $t \mapsto B(\alpha, \beta; t)$ is obviously increasing, and hence

$$
\lambda \mapsto B\left(\alpha, \beta; \frac{m^{\alpha_2-\alpha_1}}{\lambda^2 + m^{\alpha_2-\alpha_1}}\right)
$$

is decreasing. Moreover, we always have $S \leq T$, so if we put

$$
\lambda = \sqrt{\frac{T}{S}},
$$

then

$$
B\left(\alpha, \beta; \frac{m^{\alpha_2-\alpha_1}}{\lambda^2 + m^{\alpha_2-\alpha_1}}\right) \leq B\left(\alpha, \beta; \frac{m^{\alpha_2-\alpha_1}}{1 + m^{\alpha_2-\alpha_1}}\right).
$$

Hence, in view of (2.21) and (2.22), it follows that

$$
\begin{aligned}
\left(\sum_{k=1}^{m} a_k\right)^{r+1} <& 2\left(\frac{1}{\alpha_2-\alpha_1} B\left(\frac{\alpha_2-r}{(\alpha_2-\alpha_1)r}, \frac{-\alpha_1+r}{(\alpha_2-\alpha_1)r}; \frac{m^{\alpha_2-\alpha_1}}{1+m^{\alpha_2-\alpha_1}}\right)\right)^r \\
&\cdot S^{\frac{\alpha_2-r}{\alpha_2-\alpha_1}} T^{\frac{-\alpha_1+r}{\alpha_2-\alpha_1}}.
\end{aligned}
$$

This yields the inequality (2.19) with C given by (2.20). □

As in Proposition 1.1, the constant given in Theorem 2.9 is not sharp for finite m.

Problem 4 Find a formula for the sharp constant

$$
C = C_{m,r,\alpha_1,\alpha_2}
$$

in the inequality (2.19).

By combining these ideas with those of Levin and Godunova, the above theorem can, in some cases, be extended to hold for more than two factors on the right-hand side of the inequality.

Theorem 2.10 Let $N \geq 2$ be an integer, and suppose that $\alpha_1 \leq \ldots \leq \alpha_N$. Moreover, let

$$r = \frac{1}{N} \sum_{j=1}^{N} \alpha_j.$$

Then, if $a_1, \ldots, a_m$ are non-negative numbers, not all zero, it holds that

$$\sum_{k=1}^{m} a_k < C \prod_{j=1}^{N} \left(\sum_{k=1}^{m} k^{\alpha_j} a_k^{r+1} \right)^{\frac{1}{N(r+1)}}. \tag{2.23}$$

We may choose

$$C = \min_{1 \leq s < N} (AN)^{\frac{1}{r+1}} B \left(\frac{s}{Nr}, \frac{N-s}{Nr}; \frac{m^{r_2 - r_1}}{\frac{s}{N-n} + m^{r_2 - r_1}} \right)^{\frac{r}{r+1}}, \tag{2.24}$$

where

$$A = \frac{s^{-\frac{s}{N}} (N-s)^{-\frac{N-s}{N}}}{(r_2 - r_1)^{\frac{1}{r}}} \tag{2.25}$$

and $r_i = r_i(s)$ are defined by

$$\frac{1}{r_1} = \frac{1}{s} \sum_{j=1}^{s} \alpha_j \quad \text{and} \quad \frac{1}{r_2} = \frac{1}{N-s} \sum_{j=s+1}^{N} \alpha_j. \tag{2.26}$$

Remark 2.15 We see that when $N = 2$, this reduces to the special case of Theorem 2.9 where $r = p$. However, for $N > 2$, the constant does not necessarily approach the sharp constant for the corresponding infinite series inequality when $m \to \infty$.

Proof of Theorem 2.10. Let

$$S_i = \sum_{k=1}^{m} k^{\alpha_i} a_k^{r+1}, \quad i = 1, \ldots, N.$$

If $\lambda_1, \ldots, \lambda_N$ are any positive numbers, we write

$$a_k = \left(\lambda_1 k^{\alpha_1} + \ldots + \lambda_N k^{\alpha_N}\right)^{-\frac{1}{r+1}} \cdot \left(\lambda_1 k^{\alpha_1} + \ldots + \lambda_N k^{\alpha_N}\right)^{\frac{1}{r+1}} a_k$$

and apply the Hölder–Rogers inequality with parameters

$$\frac{r+1}{r} \quad \text{and} \quad r+1,$$

which yields

$$\left(\sum_{k=1}^{m} a_k\right)^{r+1} < \left(\sum_{k=1}^{m} \left(\lambda_1 k^{\alpha_1} + \ldots + \lambda_N k^{\alpha_N}\right)^{-\frac{1}{r}}\right)^{r} \sum_{i=1}^{N} \lambda_i S_i.$$

The first sum on the right-hand side can be estimated by the integral

$$I = \int_0^m \left(\lambda_1 x^{\alpha_1} + \ldots + \lambda_N x^{\alpha_N}\right)^{-\frac{1}{r}} dx.$$

Let $1 \leq s < N$, and let r_1 and r_2 be as defined by (2.26). Moreover, put

$$D_1 = s(\lambda_1 \cdots \lambda_s)^{\frac{1}{s}} \quad \text{and} \quad D_2 = (N-s)(\lambda_{s+1} \cdots \lambda_N)^{\frac{1}{N-s}}.$$

By the arithmetic–geometric mean inequality, it holds that

$$\begin{aligned}
&\lambda_1 x^{\alpha_1} + \ldots + \lambda_N x^{\alpha_N} \\
=&\left(\lambda_1 x^{\alpha_1} + \ldots + \lambda_s x^{\alpha_s}\right) + \left(\lambda_{s+1} x^{\alpha_{s+1}} + \ldots + \lambda_N x^{\alpha_N}\right) \\
=&s\frac{\lambda_1 x^{\alpha_1} + \ldots + \lambda_s x^{\alpha_s}}{s} + (N-s)\frac{\lambda_{s+1} x^{\alpha_{s+1}} + \ldots + \lambda_N x^{\alpha_N}}{N-s} \\
\geq&s\left(\lambda_1 x^{\alpha_1} \cdots \lambda_s x^{\alpha_s}\right)^{\frac{1}{s}} + (N-s)\left(\lambda_{s+1} x^{\alpha_{s+1}} \cdots \lambda_N x^{\alpha_N}\right)^{\frac{1}{N-s}} \\
=&D_1 x^{r_1} + D_2 x^{r_2},
\end{aligned}$$

and therefore, also using the relations

$$\frac{r_2 - r}{r_2 - r_1} = \frac{s}{N} \quad \text{and} \quad \frac{r - r_1}{r_2 - r_1} = \frac{N-s}{N}$$

we get

$$\begin{aligned}
I \leq& \int_0^m \left(D_1 x^{r_1} + D_2 x^{r_2}\right)^{-\frac{1}{r}} dx \\
=& \frac{\left(D_1^{-(r_2-r)} D_2^{-(r-r_1)}\right)^{\frac{1}{r(r_2-r_1)}}}{r_2 - r_1} B\left(\frac{r_2 - r}{r(r_2-r_1)}, \frac{r - r_1}{r(r_2 - r_1)}; \frac{m^{r_2-r_1}}{\frac{D_1}{D_2} + m^{r_2-r_1}}\right) \\
=& \frac{\left(D_1^{-\frac{s}{N}} D_2^{-\frac{N-s}{N}}\right)^{\frac{1}{r}}}{r_2 - r_1} B\left(\frac{s}{Nr}, \frac{N-s}{Nr}; \frac{m^{r_2-r_1}}{\frac{D_1}{D_2} + m^{r_2-r_1}}\right) \\
=& \frac{s^{-\frac{s}{Nr}} (N-s)^{-\frac{N-s}{Nr}} (\lambda_1 \cdots \lambda_N)^{-\frac{1}{Nr}}}{r_2 - r_1} B\left(\frac{s}{Nr}, \frac{N-s}{Nr}; \frac{m^{r_2-r_1}}{\frac{D_1}{D_2} + m^{r_2-r_1}}\right).
\end{aligned}$$

Now, choose the numbers λ_j so that

$$\lambda_j S_j = \left(\prod_{i=1}^{N} S_i\right)^{\frac{1}{N}}, \quad j = 1, \ldots, N. \tag{2.27}$$

Thus

$$\lambda_1 \cdots \lambda_N = 1,$$

and if A is defined by (2.25) it follows that

$$\left(\sum_{k=1}^{m} a_k\right)^{r+1} < AN \cdot B\left(\frac{s}{Nr}, \frac{N-s}{Nr}; \frac{m^{r_2-r_1}}{\frac{D_1}{D_2} + m^{r_2-r_1}}\right)^r \left(\prod_{i=1}^{N} S_i\right)^{\frac{1}{N}}.$$

Now, we note that

$$\frac{D_1}{D_2} = \frac{s}{N-s}(\lambda_1 \cdots \lambda_s)^{\frac{N}{s(N-s)}}.$$

Since S_j increases with j, it follows from (2.27) that λ_j decreases with j, and since the product of all of them equals 1, it must hold that

$$\lambda_1 \cdots \lambda_s \geq 1.$$

Thus

$$B\left(\cdot, \cdot; \frac{m^{r_2-r_1}}{\frac{D_1}{D_2} + m^{r_2-r_1}}\right) \leq B\left(\cdot, \cdot; \frac{m^{r_2-r_1}}{\frac{s}{N-s} + m^{r_2-r_1}}\right).$$

Since this can be done for any s, the conclusion follows upon taking $(r+1)$th roots. □

As was mentioned prior to Theorem 2.10, the constant is not sharp when $N > 2$, even if we let m tend to infinity.

Problem 5 Find the sharp constant

$$C = C_{m,r,N,\alpha_1,\ldots,\alpha_N}$$

in the inequality (2.23).

Remark 2.16 There are also continuous versions of the above two theorems, i.e. where the sums are replaced by integrals over bounded intervals. This will be discussed in Chapter 3.

Chapter 3

The Continuous Case

Implicit in Carlson's original paper [23], we find the following integral version of Theorem 1.1.

Theorem 3.1 If f is a non-negative, measurable function on $(0, \infty)$, then

$$\left(\int_0^\infty f(x)\,dx\right)^4 \leq \pi^2 \int_0^\infty f^2(x)\,dx \int_0^\infty x^2 f^2(x)\,dx. \tag{3.1}$$

Here, as before, the constant π^2 is the best possible. Equality in (3.1) is acheived precisely when f has the form

$$f(x) = \frac{a}{1 + bx^2},$$

where a and b are constants.

Below we give some different proofs of this theorem. We will also show how the discrete inequality (1.1) follows from the continuous version (3.1). We thus have some alternate proofs of (1.1), in addition to those presented in Section 1.3 of Chapter 1.

Our first proof of Theorem 3.1 is analogous to Hardy's first proof of (1.1).

Proof. Put

$$U = \int_0^\infty f^2(x)\,dx \quad \text{and} \quad V = \int_0^\infty x^2 f^2(x)\,dx.$$

We may assume that both U and V are finite. Let α and β be positive numbers. It then follows by the Schwarz inequality that

$$\begin{aligned}
\left(\int_0^\infty f(x)\,dx\right)^2 &= \left(\int_0^\infty \frac{1}{\sqrt{\alpha+\beta x^2}}\sqrt{\alpha+\beta x^2} f(x)\,dx\right)^2 \\
&\le \int_0^\infty \frac{dx}{\alpha+\beta x^2}\left\{\alpha\int_0^\infty f^2(x)\,dx + \beta\int_0^\infty x^2 f^2(x)\,dx\right\} \\
&= \frac{\pi}{2}\frac{1}{\sqrt{\alpha\beta}}(\alpha U + \beta V) \\
&= \frac{\pi}{2}\left(\sqrt{\frac{\alpha}{\beta}}U + \sqrt{\frac{\beta}{\alpha}}V\right).
\end{aligned}$$

With $\alpha = V$ and $\beta = U$, the rightmost expression is

$$\pi\sqrt{UV},$$

which, upon squaring both sides, yields the desired inequality. Equality in the application of the Schwarz inequality is obtained precisely when for almost all x

$$f(x) = \frac{A}{\alpha+\beta x^2}$$

for some constant A or

$$f(x) = \frac{a}{1+bx^2},$$

as claimed. □

Our second proof is based on Calculus of Variations. It is, in fact, somewhat similar to Carlson's original proof of (1.1).

Proof by Calculus of Variations. Suppose that

$$\int_0^\infty f^2(x)\,dx = a^2 \quad \text{and} \quad \int_0^\infty x^2 f^2(x)\,dx = b^2, \tag{3.2}$$

where a and b are some non-zero numbers. We want to find the infimum of

$$-\int_0^\infty f(x)\,dx$$

subject to the constraints (3.2). The Lagrange function for this problem is given by

$$L(\lambda_0,\lambda_1,\lambda_2) = \int_0^\infty \Big[-\lambda_0 f(x) + \lambda_1 f^2(x) + \lambda_2 x^2 f^2(x)\Big]\,dx. \tag{3.3}$$

Let us denote by $\bar{f}$ the minimizer of the integrand in (3.3). If $\lambda_0 = 0$, we must have $\bar{f} \equiv 0$ which cannot satisfy (3.2). By dividing through by λ_0 we may therefore assume that $\lambda_0 = 1$. Thus, let us consider

$$h(u) = -u + \lambda_1 u^2 + \lambda_2 x^2 u^2.$$

Then

$$h'(u) = -1 + 2(\lambda_1 + \lambda_2 x^2)u,$$

so that

$$h'(u) = 0$$

when

$$u = \frac{1}{2(\lambda_1 + \lambda_2 x^2)} = \bar{f}(x).$$

We have

$$\begin{aligned}\int_0^\infty \bar{f}(x)\,dx &= \frac{1}{2}\int_0^\infty \frac{dx}{\lambda_1 + \lambda_2 x^2}\\ &= \frac{\pi}{4\sqrt{\lambda_1\lambda_2}},\end{aligned}$$

$$\begin{aligned}\int_0^\infty \bar{f}^2(x)\,dx &= \frac{1}{4}\int_0^\infty \frac{dx}{(\lambda_1 + \lambda_2 x^2)^2}\\ &= \frac{\pi}{16\lambda_1\sqrt{\lambda_1\lambda_2}},\end{aligned}$$

and

$$\begin{aligned}\int_0^\infty x^2\bar{f}^2(x)\,dx &= \frac{1}{4}\int_0^\infty \frac{x^2\,dx}{(\lambda_1 + \lambda_2 x^2)^2}\\ &= \frac{\pi}{16\lambda_2\sqrt{\lambda_1\lambda_2}}.\end{aligned}$$

Then, in terms of a and b, we get

$$\bar{f}(x) = \frac{2}{\pi}\frac{a^{3/2}b^{3/2}}{b^2 + a^2x^2}.$$

It follows that the sought constant is given by

$$C = \left(\int_0^\infty \bar{f}(x)\,dx\right)^4 \bigg/ \int_0^\infty \bar{f}^2(x)\,dx \int_0^\infty x^2 \bar{f}^2(x)\,dx$$
$$= \frac{(\sqrt{ab})^4}{\frac{a^2}{\pi}\frac{b^2}{\pi}} = \pi^2.$$

□

Remark 3.1 The main advantage of the method used in the above proof is that it automatically gives the sharp constant and extremizing functions. We get this, however, on the expense of more tedious calculations. B. Kjellberg [45] mentioned that the corresponding calculations can be made for a more general case (see Chapter 4).

R. Bellman [9] proved, in addition to Theorem 2.4, the corresponding continuous versions (see Theorem 3.2 below). Bellman's basic trick was to split the integral on the left-hand side into a sum of two integrals. By streamlining his proof, using specific values of the parameters, we arrive at the following proof of (3.1) with π^2 replaced by 16.

Proof by "Partition of Unity" 1. Since

$$1 = \frac{1}{1+y} + \frac{y}{1+y},$$

it follows by the Schwarz inequality that if g is any positive function, then

$$\int_0^\infty g(y)\,dy = \int_0^\infty \frac{1}{1+y} g(y)\,dy + \int_0^\infty \frac{1}{1+y} y g(y)\,dy$$
$$\le \left(\int_0^\infty \frac{1}{(1+y)^2}\,dy\right)^{\frac{1}{2}} \left(\int_0^\infty g^2(y)\,dy\right)^{\frac{1}{2}}$$
$$+ \left(\int_0^\infty \frac{1}{(1+y)^2}\,dy\right)^{\frac{1}{2}} \left(\int_0^\infty y^2 g^2(y)\,dy\right)^{\frac{1}{2}}$$
$$= \left(\int_0^\infty g^2(y)\,dy\right)^{\frac{1}{2}} + \left(\int_0^\infty y^2 g^2(y)\,dy\right)^{\frac{1}{2}}.$$

Let δ be any positive number, and put $g(y) = f(y/\delta)$ above. After substituting x for y/δ, this yields

$$\delta \int_0^\infty f(x)\,dx \le \left(\delta \int_0^\infty f^2(x)\,dx\right)^{\frac{1}{2}} + \left(\delta^3 \int_0^\infty x^2 f^2(x)\,dx\right)^{\frac{1}{2}}$$

or

$$\int_0^\infty f(x)\,dx \le \delta^{-\frac{1}{2}} \left(\int_0^\infty f^2(x)\,dx \right)^{\frac{1}{2}} + \delta^{\frac{1}{2}} \left(\int_0^\infty x^2 f^2(x)\,dx \right)^{\frac{1}{2}}.$$

If we put

$$\delta = \left(\frac{\int_0^\infty f^2(x)\,dx}{\int_0^\infty x^2 f^2(x)\,dx} \right)^{\frac{1}{2}},$$

we get

$$\int_0^\infty f(x)\,dx \le 2 \left(\int_0^\infty f^2(x)\,dx \right)^{\frac{1}{4}} \left(\int_0^\infty x^2 f^2(x)\,dx \right)^{\frac{1}{4}},$$

which, after taking 4th powers, is (3.1) with 16 in place of π^2. □

Another way to write the constant function 1 as a sum of two functions is used in the following proof.

Proof by "Partition of Unity" 2. Let δ be any positive number. The Schwarz inequality then implies that

$$\begin{aligned}
\int_0^\infty f(x)\,dx &= \int_0^\delta f(x)\,dx + \int_\delta^\infty f(x)\,dx \\
&= \int_0^\delta 1 \cdot f(x)\,dx + \int_\delta^\infty \frac{1}{x} x f(x)\,dx \\
&\le \left(\int_0^\delta dx \right)^{\frac{1}{2}} \left(\int_0^\delta f^2(x)\,dx \right)^{\frac{1}{2}} \\
&\quad + \left(\int_\delta^\infty \frac{1}{x^2}\,dx \right)^{\frac{1}{2}} \left(\int_\delta^\infty x^2 f^2(x)\,dx \right)^{\frac{1}{2}} \\
&\le \delta^{\frac{1}{2}} \left(\int_0^\infty f^2(x)\,dx \right)^{\frac{1}{2}} + \delta^{-\frac{1}{2}} \left(\int_0^\infty x^2 f^2(x)\,dx \right)^{\frac{1}{2}}.
\end{aligned}$$

With

$$\delta = \left(\frac{\int_0^\infty x^2 f^2(x)\,dx}{\int_0^\infty f^2(x)\,dx} \right)^{\frac{1}{2}},$$

also taking 4th powers, we arrive at

$$\left(\int_0^\infty f(x)\,dx \right)^4 \le 16 \int_0^\infty f^2(x)\,dx \int_0^\infty x^2 f^2(x)\,dx,$$

i.e. the same conclusion as in the previous proof. □

Remark 3.2 The two preceding proofs have not given the best constant in (3.1), as π^2 is replaced by 16. However, the best constant can be achieved also with this idea of proof. This fact is shown in our next example, which can also be seen as a more general and unified approach to the idea used in the two proofs above.

Proof by "Partition of Unity" 3. Let k be any measurable function on $(0,\infty)$ taking values in $[0,1]$. Then, since $1 = (1-k(x)) + k(x)$, it follows by the Schwarz inequality that

$$\begin{aligned}\int_0^\infty f(x)\,dx &= \int_0^\infty (1-k(x))f(x)\,dx + \int_0^\infty \frac{k(x)}{x} x f(x)\,dx \\ &\le \left(\int_0^\infty (1-k(x))^2\,dx\right)^{\frac{1}{2}} \left(\int_0^\infty f^2(x)\,dx\right)^{\frac{1}{2}} \\ &\quad + \left(\int_0^\infty \left(\frac{k(x)}{x}\right)^2 dx\right)^{\frac{1}{2}} \left(\int_0^\infty x^2 f^2(x)\,dx\right)^{\frac{1}{2}}.\end{aligned}$$

From here, we may proceed e.g. as in any of the following three cases.

(1) Replace $f(x)$ by $f(x/\delta)$ and put $x = \delta y$ in the integrals involving f. Rename y as x and divide through by δ. This yields

$$\begin{aligned}\int_0^\infty f(x)\,dx \le &\delta^{-\frac{1}{2}} \left(\int_0^\infty (1-k(x))^2\,dx\right)^{\frac{1}{2}} \left(\int_0^\infty f^2(x)\,dx\right)^{\frac{1}{2}} \\ &+\delta^{\frac{1}{2}} \left(\int_0^\infty \left(\frac{k(x)}{x}\right)^2 dx\right)^{\frac{1}{2}} \left(\int_0^\infty x^2 f^2(x)\,dx\right)^{\frac{1}{2}}.\end{aligned}$$

If we choose

$$\delta = \left(\frac{\int_0^\infty (1-k(x))^2\,dx \int_0^\infty f^2(x)\,dx}{\int_0^\infty (k(x)/x)^2\,dx \int_0^\infty x^2 f^2(x)\,dx}\right)^{\frac{1}{2}}$$

and take 4th powers, we get

$$\begin{aligned}\left(\int_0^\infty f(x)\,dx\right)^4 \le 16 \int_0^\infty (1-k(x))^2\,dx \int_0^\infty \left(\frac{k(x)}{x}\right)^2 dx \\ \times \int_0^\infty f^2(x)\,dx \int_0^\infty x^2 f^2(x)\,dx.\end{aligned}$$

We are now free to make any choice of the function k of x. If we let

$$k(x) = \frac{x}{1+x},$$

then the two integrals involving $k(x)$ both evaluate to 1, and we essentially have the first "Partition of Unity" proof above.

(2) Let

$$k = \chi_{(\delta^{-1},\infty)}.$$

Then

$$\int_0^\infty (1-k(x))^2\,dx = \delta^{-1}$$

and

$$\int_0^\infty \left(\frac{k(x)}{x}\right)^2 dx = \delta.$$

With

$$\delta = \left(\frac{\int_0^\infty f^2(x)\,dx}{\int_0^\infty x^2 f^2(x)\,dx}\right)^{\frac{1}{2}},$$

this is essentially the second "Partition of Unity" proof. Hence, after taking 4th powers, we get the desired inequality with constant 16.

(3) In the two applications of the Schwarz inequality, we obtain *equality* if for some constants λ and μ we have

$$f^2(x) = \lambda(1-k(x))^2$$

and

$$x^2 f^2(x) = \mu\left(\frac{k(x)}{x}\right)^2,$$

respectively. Solving for k as a function of x yields

$$k(x) = \frac{\delta^2 x^2}{1+\delta^2 x^2}$$

for some $\delta > 0$. Thus

$$\int_0^\infty (1-k(x))^2\,dx = \int_0^\infty \frac{1}{(1+\delta x^2)^2}\,dx = \delta^{-1}\frac{\pi}{4}$$

and

$$\int_0^\infty \left(\frac{k(x)}{x}\right)^2 dx = \int_0^\infty \frac{\delta^2 x^2}{(1+\delta x^2)^2}\, dx = \delta \frac{\pi}{4}.$$

If we put

$$\delta = \left(\frac{\int_0^\infty f^2(x)\, dx}{\int_0^\infty x^2 f^2(x)\, dx}\right)^{\frac{1}{2}},$$

this yields

$$\int_0^\infty f(x)\, dx \le 2\sqrt{\frac{\pi}{4}} \left(\int_0^\infty f^2(x)\, dx\right)^{\frac{1}{4}} \left(\int_0^\infty x^2 f^2(x)\, dx\right)^{\frac{1}{4}},$$

or

$$\left(\int_0^\infty f(x)\, dx\right)^4 \le \pi^2 \int_0^\infty f^2(x)\, dx \int_0^\infty x^2 f^2(x)\, dx.$$

Thus, in the last case, the function k is chosen optimally. Moreover, any case of equality can be traced, by solving for $f(x)$ once $k(x)$ is determined. We find that equality occurs if and only if f has the form

$$f(x) = \frac{\gamma}{1+\delta^2 x^2}.$$

□

We conclude this section by proving that the continuous version of Carlson's inequality indeed implies the discrete version, thus obtaining some new alternate ways of proving the discrete inequality.

Proposition 3.1 The discrete inequality (1.1) follows from its integral analogue (3.1).

Proof. Suppose that (3.1) holds, and let $\{a_k\}_{k=1}^\infty$ be a non-zero sequence of non-negative numbers. Fix the positive integer m for the moment, and define

$$f_m(x) = \sum_{k=1}^m a_k \chi_{[k-1,k)}(x),$$

letting χ_I denote the characteristic function of the interval I. Since at least one a_k is non-zero, the function f_m is non-zero for m sufficiently large. Thus, for such m, it is not of the form

$$f(x) = \frac{a}{1+bx^2}$$

for any constants a and b, and hence we have strict inequality in (3.1) for this function. Since

$$\chi_{[k-1,k)}(x)\chi_{[j-1,j)}(x) = 0$$

if $k \neq j$ and, moreover, $\chi_I^2 = \chi_I$ for any interval I, it follows that

$$\begin{aligned}
\left(\sum_{k=1}^{m} a_k\right)^4 &= \left(\sum_{k=1}^{m} a_k \int_0^\infty \chi_{[k-1,k)}\,dx\right)^4 \\
&= \left(\int_0^\infty f_m(x)\,dx\right)^4 \\
&< \pi^2 \int_0^\infty f_m^2(x)\,dx \int_0^\infty x^2 f_m^2(x)\,dx \\
&= \pi^2 \sum_{k=1}^{m} a_k^2 \int_{k-1}^{k} dx \sum_{k=1}^{m} a_k^2 \int_{k-1}^{k} x^2\,dx \\
&\leq \pi^2 \sum_{k=1}^{m} a_k^2 \sum_{k=1}^{m} k^2 a_k^2.
\end{aligned}$$

This result follows by letting $m \to \infty$. □

3.1 Beurling

In 1938, when studying Fourier transforms, A. Beurling [15] proved that

$$\frac{1}{\sqrt{2\pi}} \int_{-\infty}^{\infty} |f(x)|\,dx \leq \left(\int_{-\infty}^{\infty} |g(t)|^2\,dt \int_{-\infty}^{\infty} |g'(t)|^2\,dt\right)^{1/4}, \tag{3.4}$$

where f is the Fourier transform of g, i.e.

$$g(t) = \frac{1}{\sqrt{2\pi}} \int_{-\infty}^{\infty} e^{itx} f(x)\,dx. \tag{3.5}$$

Another application of this to Fourier transforms can be found in P. Brenner and V. Thomée [16].

Remark 3.3 The remark on absolute values when considering infinite series also applies in the continuous case, i.e. we may put absolute values on the functions appearing in the integrals, thus also covering the case of complex-valued functions. In Beurling's inequality, however, the absolute values are necessary, since the Fourier transform of f may not be real-valued even though f is.

Beurling used the same ideas as Hardy for his proof. By applying the Parseval identity to (3.4), we arrive at an inequality similar to (3.1), namely

$$\int_{-\infty}^{\infty} |f(x)|\,dx \leq \sqrt{2\pi}\left(\int_{-\infty}^{\infty} |f(x)|^2\,dx \int_{-\infty}^{\infty} x^2|f(x)|^2\,dx\right)^{1/4}. \tag{3.6}$$

In fact, (3.1) and (3.6) are equivalent. To see this, note that

$$\begin{aligned} a_0^{1/4}b_0^{1/4} + a_1^{1/4}b_1^{1/4} &= \sum_{i=0}^{1} 1^{1/2}a_i^{1/4}b_i^{1/4} \\ &\leq \left(\sum_{i=0}^{1} 1\right)^{1/2}\left(\sum_{i=0}^{1} a_i\right)^{1/4}\left(\sum_{i=0}^{1} b_i\right)^{1/4} \\ &= \sqrt{2}(a_0+a_1)^{1/4}(b_0+b_1)^{1/4}. \end{aligned} \tag{3.7}$$

Assume first that (3.1) holds. It is clear that we can apply Carlson's inequality to the functions f_0 and f_1, defined by

$$\begin{aligned} f_0(x) &= f(x), \quad x > 0, \\ f_1(x) &= f(-x), \quad x < 0, \end{aligned}$$

which yields

$$\begin{aligned} \int_{-\infty}^{\infty} f(x)\,dx \leq \sqrt{\pi}\Bigg\{&\left(\int_{-\infty}^{0} f^2(x)\,dx\right)^{1/4}\left(\int_{-\infty}^{0} x^2 f^2(x)\,dx\right)^{1/4} \\ &+ \left(\int_{0}^{\infty} f^2(x)\,dx\right)^{1/4}\left(\int_{0}^{\infty} x^2 f^2(x)\,dx\right)^{1/4}\Bigg\}, \end{aligned}$$

so that by (3.7) with the proper choices of a_i and b_i we get (3.6). Assume, conversely, that (3.6) holds. Any function defined on $(0,\infty)$ can be extended to, say, an even function on $(-\infty,\infty)$, and the value of the integral of the extended function is then, of course, twice the value of the integral over

$(0, \infty)$. To any such function, (3.6) can be applied, and we get

$$\begin{aligned}
\int_0^\infty f(x)\,dx &= \frac{1}{2}\int_{-\infty}^\infty f(x)\,dx \\
&\le \frac{1}{2}\sqrt{2\pi}\left(\int_{-\infty}^\infty f^2(x)\,dx\right)^{1/4}\left(\int_{-\infty}^\infty x^2 f^2(x)\,dx\right)^{1/4} \\
&= \sqrt{\frac{\pi}{2}}\left(2\int_0^\infty f^2(x)\,dx\right)^{1/4}\left(2\int_0^\infty x^2 f^2(x)\,dx\right)^{1/4} \\
&= \sqrt{\pi}\left(\int_0^\infty f^2(x)\,dx\right)^{1/4}\left(\int_0^\infty x^2 f^2(x)\,dx\right)^{1/4}.
\end{aligned}$$

In Remark 4.2 of Chapter 4, a more general version of this equivalence is pointed out.

Remark 3.4 Note that the inequality (3.4) can be thought of as an uncertainty inequality.

3.2 Kjellberg

B. Kjellberg [44] published in 1946 a paper with a slightly different viewpoint. Below, the main thoughts from this paper are reconstructed.

Consider Carlson's inequality (1.1) and its continuous counterpart (3.1). They state that if the series and integrals, respectively, on the right-hand sides are finite, then so are those on the left-hand sides. This is the essence of Kjellberg's way of looking at the inequalities. Note, however, that in (1.1), the finiteness of the series $\sum a_k^2$ is implied by that of the series $\sum k^2 a_k^2$, while in (3.1) the convergence of $\int f^2(x)\,dx$ and $\int x^2 f^2(x)\,dx$ are of equal importance when asserting the convergence of $\int f(x)\,dx$. Furthermore, if the finiteness of $\sum k^2 a_k^2$ is known, then directly by the Schwarz inequality

$$\begin{aligned}
\sum_{k=1}^\infty a_k &= \sum_{k=1}^\infty \frac{1}{k} k a_k \\
&\le \sqrt{\sum_{k=1}^\infty \frac{1}{k^2}\sum_{k=1}^\infty k^2 a_k^2} \qquad (3.8)\\
&= \frac{\pi}{\sqrt{6}}\left(\sum_{k=1}^\infty k^2 a_k^2\right)^{1/2}.
\end{aligned}$$

Now, since

$$\frac{\pi}{\sqrt{6}} < \sqrt{\pi},$$

the inequality (3.8) may be numerically better than (1.1). For these reasons, Kjellberg found it more interesting to study the continuous case. The basic Kjellberg idea is as follows.

If α and β are real numbers, let

$$I(\alpha,\beta) = \int_0^\infty x^\alpha f^\beta(x)\,dx.$$

Thus, for instance, we can rewrite (3.1) as

$$I(0,1)^4 \leq \pi^2 I(0,2) I(2,2).$$

Moreover, if

$$\begin{aligned} \alpha &= (1-\theta)\alpha_1 + \theta\alpha_2 \\ \text{and} \quad \beta &= (1-\theta)\beta_1 + \theta\beta_2 \end{aligned}$$

for some $\theta \in (0,1)$, i.e. if the point (α,β) is situated on the straight line segment between the two points (α_1,β_1) and (α_2,β_2) in the $\alpha\beta$-plane, then, by applying the Hölder–Rogers inequality with the conjugate exponents

$$\frac{1}{1-\theta} \quad \text{and} \quad \frac{1}{\theta},$$

we arrive at

$$I(\alpha,\beta) \leq I(\alpha_1,\beta_1)^{1-\theta} I(\alpha_2,\beta_2)^\theta.$$

Thus, if the integrals $I(\alpha_1,\beta_1)$ and $I(\alpha_2,\beta_2)$ converge, then so does the integral corresponding to any point on the line segment joining the two points (α_1,β_1) and (α_2,β_2). More generally, we have the following principle, illustrated in Figure 3.1 (for the case $m=4$).

Proposition 3.2 (The Kjellberg Principle) Suppose that $m \geq 2$,

$$I(\alpha_j,\beta_j) < \infty, \quad j = 1,\ldots,m$$

and that (α,β) is a point in the convex hull of the points $(\alpha_1,\beta_1),\ldots,(\alpha_m,\beta_m)$. Then

$$I(\alpha,\beta) < \infty.$$

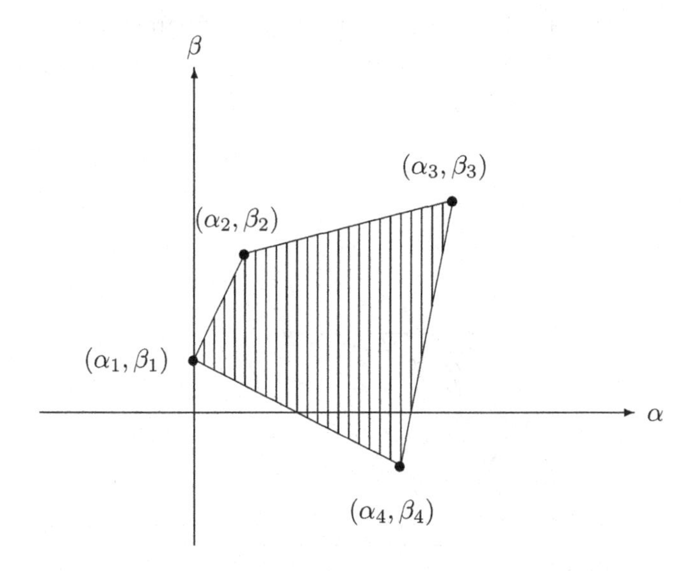

Fig. 3.1 If the integrals corresponding to the points (α_j, β_j), $j = 1, 2, 3, 4$ converge, then so does the integral corresponding to each point in their convex hull.

Example 3.1 Assume that

$$\int_0^\infty f^2(x)\,dx < \infty \quad \text{and}$$

$$\int_0^\infty x^2 f^2(x)\,dx < \infty,$$

as suggested by the inequality (3.1). Suppose that $0 < T < \infty$ and write

$$\int_0^\infty x^\alpha f^\beta(x)\,dx = \int_0^T x^\alpha f^\beta(x)\,dx + \int_T^\infty x^\alpha f^\beta(x)\,dx.$$

By assumption, the two integrals on the right-hand side are finite when $\beta = 2$ and $\alpha = 0, 2$. Thus, by the Kjellberg Principle, also for $\beta = 2$ and α any number between 0 and 2. To see this, consider functions f which vanish off $(0, T)$ and off (T, ∞), respectively. Furthermore, for any $\alpha > -1$

$$\int_0^T x^\alpha f^0(x)\,dx < \infty,$$

so again by the Kjellberg Principle

$$\int_0^T x^\alpha f^\beta(x)\,dx < \infty$$

whenever (α, β) is in the shaded region of the upper diagram in Figure 3.2.

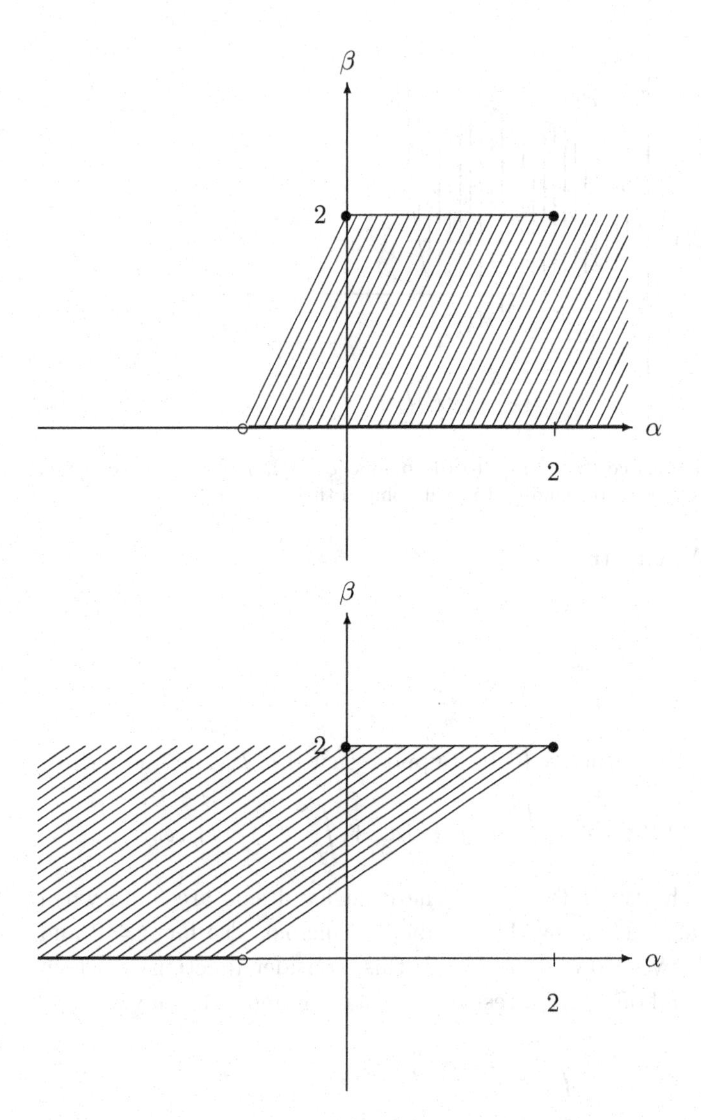

Fig. 3.2 The shaded region in the upper diagram represents the points (α, β) for which $\int_0^T x^\alpha f^\beta(x)\,dx$ is finite, under the assumption that $\int_0^T f^2(x)\,dx$ and $\int_0^T x^2 f^2(x)\,dx$ are both finite. The shaded region in the lower diagram represents the corresponding points for integrals over (T, ∞).

Similarly, since

$$\int_T^\infty x^\alpha f^0(x)\,dx < \infty$$

whenever $\alpha < -1$, our assumptions and the Kjellberg principle imply that

$$\int_T^\infty x^\alpha f^\beta(x)\,dx < \infty$$

for all (α, β) in the lower shaded region in Figure 3.2. Together, these two observations imply that

$$\int_0^\infty x^\alpha f^\beta(x)\,dx < \infty$$

for all (α, β) in the region shown in Figure 3.3. Note, in particular, that the point $(0, 1)$, corresponding to the integral on the left-hand side of (3.1), is contained in this region.

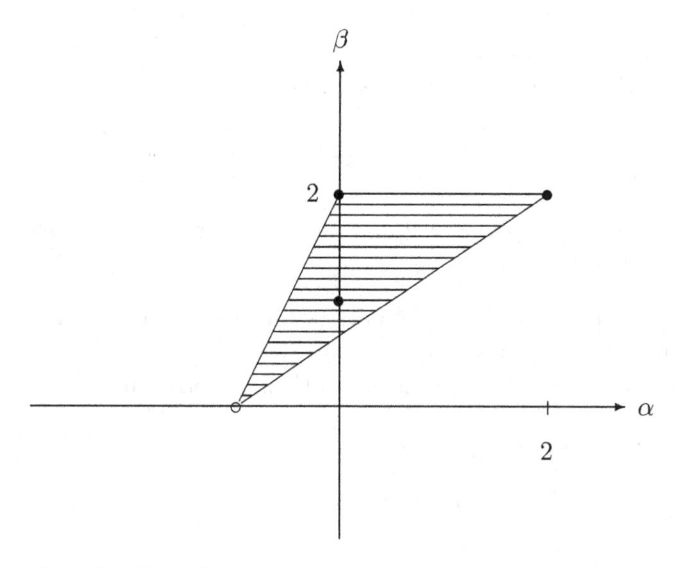

Fig. 3.3 The point corresponding to the integral on the left-hand side of Carlson's integral inequality is in the region of convergence, provided that the integrals on the right-hand side of the same inequality converge.

It is clear that this method can be extended to involve any finite number of points corresponding to convergent integrals. Although we lose track of best constants when forming inequalities of the form (3.1), the method is indeed very powerful, and the geometric nature gives intuition a great deal of help when dealing with problems of this kind.

In Chapter 5, some more of Kjellberg's results will be discussed.

3.3 Bellman

Theorem 2.4 was shown by Bellman [9] to hold also for integrals.

Theorem 3.2 (Bellman, 1943)

(a) Suppose that $p, q > 1$ and $\lambda, \mu > 0$. Then there is a constant C such that

$$\left(\int_0^\infty f(x)\,dx\right)^{p\mu+q\lambda} \le C\left(\int_0^\infty x^{p-1-\lambda} f^p(x)\,dx\right)^{\mu} \left(\int_0^\infty x^{q-1+\mu} f^q(x)\,dx\right)^{\lambda} \tag{3.9}$$

holds for all non-negative functions f.

(b) If $\alpha, \beta > 1$, then there is a constant C such that

$$\left(\int_0^\infty f(x)\,dx\right)^{\alpha\beta+\alpha-\beta} \le C\int_0^\infty f^\alpha(x)\,dx\left(\int_0^\infty x^\beta f^\beta(x)\,dx\right)^{\alpha-1}.$$

Remark 3.5 Obviously, with $p = q = 2$ and $\lambda = \mu = 1$ in part (a), and with $\alpha = \beta = 2$ in part (b), we get (3.1).

We will reconstruct Bellman's proof of part (a) of this theorem, which uses the next lemma. However, in order to be able to track the constant in (3.9), we state the lemma in a more precise form than the original one.

Lemma 3.1 If $a, b, c > 0$ and $v > u > 0$ are such that

$$b\rho^u \le c + a\rho^v, \quad \rho > 0,$$

then

$$c^{v-u}a^u \ge \frac{u^u(v-u)^{v-u}}{v^v} b^v.$$

Proof. The statement follows if we put

$$\rho = \left(\frac{bu}{av}\right)^{1/(v-u)}.$$

□

***Proof of Theorem 3.2* (a).** Note that for any r, s we can write

$$1 = \frac{x^{r/p}}{x^{r/p}(1+x)} + \frac{x^{s/q}}{x^{s/q}\left(1+\frac{1}{x}\right)}.$$

Thus, by applying the Hölder–Rogers inequality and letting $r = p-1-\lambda$, $s = q-1+\mu$, we find that

$$\begin{aligned}
\int_0^\infty f(x)\,dx &= \int_0^\infty \frac{x^{r/p}}{x^{r/p}(1+x)} f(x)\,dx + \int_0^\infty \frac{x^{s/q}}{x^{s/q}\left(1+\frac{1}{x}\right)} f(x)\,dx \\
&\le \left(\int_0^\infty \frac{dx}{x^{r/(p-1)}(1+x)^{p'}}\right)^{1/p'} \left(\int_0^\infty x^r f^p(x)\,dx\right)^{1/p} \\
&+ \left(\int_0^\infty \frac{dx}{x^{s/(q-1)}\left(1+\frac{1}{x}\right)^{q'}}\right)^{1/q'} \left(\int_0^\infty x^s f^q(x)\,dx\right)^{1/q} \\
&= B\left(\frac{\lambda}{p-1}, \frac{p-\lambda}{p-1}\right)^{1/p'} \left(\int_0^\infty x^{p-1-\lambda} f^p(x)\,dx\right)^{1/p} \\
&+ B\left(\frac{\mu}{q-1}, \frac{q-\mu}{q-1}\right)^{1/q'} \left(\int_0^\infty x^{q-1+\mu} f^q(x)\,dx\right)^{1/q} \\
&=: L\left(\int_0^\infty x^{p-1-\lambda} f^p(x)\,dx\right)^{1/p} + M\left(\int_0^\infty x^{q-1+\mu} f^q(x)\,dx\right)^{1/q}.
\end{aligned}$$

Replacing $f(x)$by $f(x/\rho)$ and making the substitution $x = \rho y$, this yields, upon switching back from y to x

$$\begin{aligned}
\rho \int_0^\infty f(x)\,dx \le & L\rho^{(p-\lambda)/p} \left(\int_0^\infty x^{p-1-\lambda} f^p(x)\,dx\right)^{1/p} \\
& + M\rho^{(q+\mu)/q} \left(\int_0^\infty x^{q-1+\mu} f^q(x)\,dx\right)^{1/q}.
\end{aligned}$$

Now, we can apply the convexity inequality $(\alpha+\beta)^t \le 2^{t-1}(\alpha^t+\beta^t)$ (which is valid for all $\alpha, \beta \ge 0$ whenever $t \ge 1$) with $t = pq$ and divide through by $\rho^{q(p-\lambda)}$, to get that

$$\left(\int_0^\infty f(x)\,dx\right)^{pq} \rho^{q\lambda} \le 2^{pq-1} L^{pq} P^q + 2^{pq-1} M^{pq} Q^p \rho^{p\mu+q\lambda},$$

where

$$P = \int_0^\infty x^{p-1-\lambda} f^p(x)\,dx \quad \text{and} \quad Q = \int_0^\infty x^{q-1+\mu} f^q(x)\,dx.$$

Thus Lemma 3.1 with

$$\begin{aligned} a &= 2^{pq-1} M^{pq} Q^p, \\ b &= \left(\int_0^\infty f(x)\,dx \right)^{pq}, \\ c &= 2^{pq-1} L^{pq} P^q, \\ u &= q\lambda \quad \text{and} \\ v &= p\mu + q\lambda \end{aligned}$$

yields

$$\left(\int_0^\infty f(x)\,dx \right)^{pq(p\mu+q\lambda)} \leq C^{pq} P^{pq\lambda} Q^{pq\mu},$$

where

$$\begin{aligned} C = & \left(\frac{(p\mu+q\lambda)^{p\mu+q\lambda}}{(p\mu)^{p\mu}(q\lambda)^{q\lambda}} \right)^{1/pq} 2^{(pq-1)(p\mu+q\lambda)/pq} \\ & B\left(\frac{\lambda}{p-1}, \frac{p-\lambda}{p-1} \right)^{(p-1)\mu} B\left(\frac{\mu}{q-1}, \frac{q-\mu}{q-1} \right)^{(q-1)\lambda}. \end{aligned} \tag{3.10}$$

The desired inequality thus follows if we take pqth roots. □

Remark 3.6 This method of proof is obviously based on one of Hardy's original ideas, but also on the genuinely new idea of proving a multiplicative inequality by going via an additive inequality. This method was later pushed to perfection by Levin [56] (see Chapter 4). It should be mentioned that the constant obtained implicitly in the above proof is not sharp, except possibly in some special cases. For instance, in the Carlson case $p = q = 2$ and $\lambda = \mu = 1$, Bellman's proof yields the constant 16 in place of π^2. The sharp constant in the general case was found by Levin. Again, we refer to Chapter 4 for details concerning this improvement.

Remark 3.7 In his book on Sobolev spaces, V. G. Maz'ja [65] stated and proved part (a) of Bellman's theorem as a lemma. This was used in connection with embeddings of Sobolev spaces.

3.4 Sz. Nagy

Beurling's inequality (3.4) and the definition (3.5) imply that for any $s \in \mathbb{R}$

$$|g(s)| \leq \left(\int_{-\infty}^{\infty} |g(t)|^2 \, dt \int_{-\infty}^{\infty} |g'(t)|^2 \, dt \right)^{1/4}, \tag{3.11}$$

so that, translated into the language of L_p-norms,

$$\|g\|_\infty \leq \|g\|_2^{1/2} \|g'\|_2^{1/2}.$$

B. Sz. Nagy [82] varied the exponent 2 in the above norms, proving in the 1941 paper the following two results, where the norms are unweighted L_p norms on $\mathbb{R}$ with Lebesgue measure. We reconstruct the proof of the first result.

Theorem 3.3 (Sz. Nagy, 1941) Suppose that $f : \mathbb{R} \to \mathbb{R}$ is absolutely continuous and let $\alpha > 0$ and $p \geq 1$. Put

$$\theta = \frac{p'}{p' + \alpha}, \tag{3.12}$$

where p' is the exponent conjugate to p:

$$\frac{1}{p} + \frac{1}{p'} = 1$$

(we put $p' = \infty$ when $p = 1$). Then

$$\|f\|_\infty \leq \frac{1}{(2\theta)^\theta} \|f\|_\alpha^{1-\theta} \|f'\|_p^\theta. \tag{3.13}$$

Remark 3.8 Note that by (3.12), we always have $0 < \theta \leq 1$. Note also that if we put $\alpha = p = 2$, then (3.13) reduces to (3.11).

Proof of Theorem 3.3. Suppose first that $p = 1$. The inequality to prove is then

$$\|f\|_\infty \leq \frac{1}{2} \|f'\|_1. \tag{3.14}$$

We may assume that the right-hand side is finite, in which case

$\lim_{x\to\pm\infty} f(x) = 0$. For any $\xi \in \mathbb{R}$

$$\begin{aligned}\int_{-\infty}^{\infty} |f'(x)|\,dx &= \int_{-\infty}^{\xi} |f'(x)|\,dx + \int_{\xi}^{\infty} |f'(x)|\,dx \\ &\geq \pm \int_{-\infty}^{\xi} f'(x)\,dx \mp \int_{\xi}^{\infty} f'(x)\,dx \\ &= \pm \lim_{T\to\infty} \left(\int_{-T}^{\xi} f'(x)\,dx + \int_{\xi}^{T} f'(x)\,dx \right) \\ &= \pm 2f(\xi).\end{aligned}$$

Thus, for all ξ, it holds that

$$|f(\xi)| \leq \frac{1}{2} \|f'\|_1 ,$$

which implies (3.14). Assume now that $p > 1$. In view of the Hölder–Rogers inequality

$$\begin{aligned}\int_{-\infty}^{\infty} |f(x)|^{\alpha/p'} |f(x)|\,dx &\leq \left(\int_{-\infty}^{\infty} |f(x)|^{\alpha}\,dx \right)^{1/p'} \left(\int_{-\infty}^{\infty} |f'(x)|^{p}\,dx \right)^{1/p} \\ &= \|f\|_{\alpha}^{(1-\theta)/\theta} \|f'\|_{p} .\end{aligned}$$

It remains to prove that

$$\int_{-\infty}^{\infty} |f(x)|^{\alpha/p'} |f'(x)|\,dx \geq 2\theta \|f\|_{\infty}^{1/\theta} . \tag{3.15}$$

It holds that

$$\begin{aligned}\int_{-\infty}^{\infty} |f(x)|^{\alpha/p'} |f'(x)|\,dx &\geq \left(\int_{-\infty}^{0} - \int_{0}^{\infty} \right) \operatorname{sgn} f(x) |f(x)|^{\alpha/p'} f'(x)\,dx \\ &= \frac{1}{\frac{\alpha}{p'}+1} \left(\int_{-\infty}^{0} - \int_{0}^{\infty} \right) \frac{d}{dx} |f(x)|^{\alpha/p'+1}\,dx \\ &= \theta \left(\int_{-\infty}^{0} - \int_{0}^{\infty} \right) \frac{d}{dx} |f(x)|^{1/\theta}\,dx \\ &= \theta \lim_{T\to\infty} \left(\int_{-T}^{0} - \int_{0}^{T} \right) \frac{d}{dx} |f(x)|^{1/\theta}\,dx \\ &= \theta \lim_{T\to\infty} [|f(0)|^{1/\theta} - |f(-T)|^{1/\theta} - |f(T)|^{1/\theta} + |f(0)|^{1/\theta}] \\ &= 2\theta |f(0)|^{1/\theta}.\end{aligned}$$

Now, it is clear that this estimate remains true if f is replaced by any translate $f_\xi(x) = f(x+\xi)$, which shows that for all ξ we have

$$2\theta |f(\xi)|^{1/\theta} \le \int_{-\infty}^{\infty} |f(x)|^{\alpha/p'} |f'(x)|\, dx,$$

and, hence, (3.15) holds and the proof is complete. □

The second result from [82], the proof of which we will omit here, allows more flexibility in parameters. If u and v are real numbers, let

$$H(u,v) = \frac{u^u v^v}{(u+v)^{u+v}} \frac{\Gamma(1+u+v)}{\Gamma(1+u)\Gamma(1+v)},$$

where $\Gamma(\cdot)$ denotes the Gamma function.

Theorem 3.4 (Sz. Nagy, 1941) Suppose that $f : \mathbb{R} \to \mathbb{R}$ is absolutely continuous and let $\alpha, \beta > 0$ and $p \ge 1$. Let

$$\eta = \frac{\beta}{\alpha+\beta} \frac{p'}{p'+\alpha}$$

and

$$q = 1 + \frac{\alpha}{p'}.$$

Then

$$\|f\|_{\alpha+\beta} \le C \, \|f\|_{\alpha}^{1-\eta} \, \|f'\|_{p}^{\eta},$$

where

$$C = \left[\frac{q}{2} H\left(\frac{q}{\beta}, \frac{1}{p'} \right) \right]^{\beta/q(\alpha+\beta)}.$$

3.5 Klefsjö

In the exercise section of Elementa **54** (1971) No. 2, B. Klefsjö [42] posed the following problem, with accompanying hint (which more or less gives the entire solution to the problem).

EXERCISE. Show that if f is a non-negative, integrable function, then

$$\int_0^\infty f(x)\, dx \le 2 \left(\int_0^\infty \sqrt{x} f(x)\, dx \right)^{\frac{1}{2}} \left(\int_0^\infty (f(x))^2\, dx \right)^{\frac{1}{4}}.$$

Hint: Estimate with the Schwarz inequality

$$\int_0^\infty g(x)h(x)\,dx \le \left(\int_0^\infty |g(x)|^2\,dx \cdot \int_0^\infty |h(x)|^2\,dx\right)^{\frac{1}{2}},$$

in easiest possible way, $\int_0^t f(x)\,dx$ and $\int_t^\infty f(x)\,dx$ by integrals on the right-hand side of the formula which is to be shown. For each $t > 0$, we have

$$\int_0^t f(x)\,dx \le \left(\int_0^t 1\,dx\right)^{\frac{1}{2}} \left(\int_0^t f^2(x)\,dx\right)^{\frac{1}{2}} \le \sqrt{t}\left(\int_0^\infty f^2(x)\,dx\right)^{\frac{1}{2}}$$

and

$$\int_t^\infty f(x)\,dx \le \frac{1}{\sqrt{t}}\int_t^\infty \sqrt{x}f(x)\,dx \le \frac{1}{\sqrt{t}}\int_0^\infty \sqrt{x}f(x)\,dx.$$

From this, it follows that

$$\begin{aligned}\int_0^\infty f(x)\,dx &= \int_0^t f(x)\,dx + \int_t^\infty f(x)\,dx \\ &\le \sqrt{t}\left(\int_0^\infty f^2(x)\,dx\right)^{\frac{1}{2}} + \frac{1}{\sqrt{t}}\int_0^\infty \sqrt{x}f(x)\,dx\end{aligned} \tag{3.16}$$

for each $t > 0$. By studying the function of t on the rightmost side of (3.16), one finds that it attains the value

$$2\left(\int_0^\infty f^2(x)\,dx\right)^{\frac{1}{4}} \left(\int_0^\infty \sqrt{x}f(x)\,dx\right)^{\frac{1}{2}}$$

for

$$t = \int_0^\infty \sqrt{x}f(x)\,dx \Big/ \left(\int_0^\infty f^2(x)\,dx\right)^{\frac{1}{2}}.$$

Remark 3.9 We leave it as a problem to the reader to examine (e.g. by comparing to other results in this book) whether the constant in Klefsjö's inequality is sharp.

3.6 Hu

In 1993, K. Hu [36] proved some multiplicative inequalities, generalizing another result by Sz. Nagy [81].

Theorem 3.5 (Hu, 1993) Suppose that $-\infty < a < b < \infty$, $\alpha > 0$ and $p > 1$, and put

$$q = 1 + \frac{\alpha}{p'}.$$

Suppose, moreover, that the function $e : [a, b] \to \mathbb{R}$ is such that

$$1 - e(x) + e(y) \geq 0, \quad x, y \in [a, b]$$

and that f is absolutely continuous with $f' \in L_1(a, b)$ and that f vanishes at some point in $[a, b]$.

(a) If $1 < p \leq 2$, then

$$|f(a)|^q + |f(b)|^q \leq q \left(\int_a^b |f'(x)|^p \, dx \right)^{2/p-1} \left[\left(\int_a^b |f'(x)|^p \, dx \int_a^b |f(x)|^\alpha \, dx \right)^2 - B^2 \right]^{1/2p'}.$$

(b) If $p > 2$, then

$$|f(a)|^q + |f(b)|^q \leq q \left(\int_a^b |f(x)|^\alpha \, dx \right)^{1-2/p} \left[\left(\int_a^b |f'(x)|^p \, dx \int_a^b |f(x)|^\alpha \, dx \right)^2 - B^2 \right]^{1/2p}.$$

In both cases, we define

$$B = \int_a^b |f'(x)|^p \, dx \int_a^b |f(x)|^\alpha e(x) \, dx - \int_a^b |f'(x)|^p e(x) \, dx \int_a^b |f(x)|^\alpha \, dx.$$

Remark 3.10 For other results regarding integrals over bounded intervals, see Section 3.9 below.

3.7 Yang–Fang

G.-S. Yang and J.-C. Fang [84] gave another generalization of Carlson's inequalities. Their paper contains the inequality in both the discrete and continuous cases. We only consider the continuous version here.

Theorem 3.6 (Yang–Fang, 1999) Let $f, g : [0, \infty) \to \mathbb{R}$ be Lebesgue measurable functions and g be continuously differentiable with $g(0) = 0$, $\lim_{x\to\infty} g(x) = \infty$ and

$$0 < M = \inf_{x\in[0,\infty)} g'(x) < \infty.$$

Suppose that p, α and r are real numbers such that $p > 2$ and $\alpha > 0$. Then

$$\left(\int_0^\infty |f(x)|\,dx\right)^{2p} \le \left(\frac{\pi}{\alpha M}\right)^2 \left(\int_0^\infty g^{1-\alpha}(x)|f(x)|^{p(1+2r-rp)}\,dx\right)$$
$$\times \left(\int_0^\infty g^{1+\alpha}(x)|f(x)|^{p(1+2r-rp)}\,dx\right)\left(\int_0^\infty |f(x)|^{rp}\,dx\right)^{2(p-2)}. \tag{3.17}$$

Remark 3.11 Theorem 3.6 generalizes a result by S. Barza and E. C. Popa [7]. See also the Ph.D. thesis [4].

A multi-dimensional extension of Theorem 3.6 will be discussed in Chapter 5.

3.8 A Continuous Landau Type Inequality

In 1998, S. Barza, J. Pečarić and L.-E. Persson [6] considered the continuous version of Landau's inequality (2.7), i.e. an inequality involving integrals with weights of the form $x - a$. We begin by presenting the following result.

Theorem 3.7 Suppose that $a > 0$. Then, for every non-negative, measurable function f on $\mathbb{R}_+$, it holds that

$$\left(\int_0^\infty f(x)\,dx\right)^4 \le 4\pi^2 \int_0^\infty f^2(x)\,dx \int_0^\infty (x-a)^2 f^2(x)\,dx, \tag{3.18}$$

and the constant $4\pi^2$ is sharp.

Remark 3.12 Note that by Carlson's Theorem 3.1, the constant in the above inequality is not sharp if we were allowed to put $a = 0$.

Proof of Theorem 3.7. We proceed as in the proof of Theorem 3.1. Thus, let

$$U = \int_0^\infty f^2(x)\,dx \quad \text{and} \quad V = \int_0^\infty (x-a)^2 f^2(x)\,dx.$$

Assume, without loss of generality, that $U < \infty$ and $V < \infty$. Let α and β be any positive numbers, and write

$$f(x) = \frac{1}{\sqrt{\alpha + \beta(x-a)^2}} \sqrt{\alpha + \beta(x-a)^2} f(x).$$

Thus by the Schwarz inequality

$$\begin{aligned}
\left(\int_0^\infty f(x)\,dx\right)^2 &= \left(\int_0^\infty \frac{1}{\sqrt{\alpha + \beta(x-a)^2}} \sqrt{\alpha + \beta(x-a)^2} f(x)\,dx\right)^2 \\
&\le \int_0^\infty \frac{dx}{\alpha + \beta(x-a)^2} \int_0^\infty (\alpha + \beta(x-a)^2) f^2(x)\,dx \\
&= \frac{1}{\sqrt{\alpha\beta}}\left(\frac{\pi}{2} + \arctan\left(a\sqrt{\frac{\beta}{\alpha}}\right)\right)(\alpha S + \beta T) \\
&= \left(\frac{\pi}{2} + \arctan\left(a\sqrt{\frac{\beta}{\alpha}}\right)\right)\left(\sqrt{\frac{\alpha}{\beta}} S + \sqrt{\frac{\beta}{\alpha}} T\right).
\end{aligned}$$

Since $\arctan t < \frac{\pi}{2}$ for any (finite) $t > 0$ (see also Remark 3.13 below), by choosing $\alpha = T$ and $\beta = S$, we obtain

$$\left(\int_0^\infty f(x)\,dx\right)^2 < \pi(\sqrt{ST} + \sqrt{ST}) = 2\pi\sqrt{ST},$$

which, after squaring, yields the desired inequality. To prove that the constant is sharp, we consider the functions

$$f_\gamma(x) = \frac{1}{1 + \gamma^2(x-a)^2}, \qquad \gamma > 0.$$

We have

$$\int_0^\infty f_\gamma(x)\,dx = \frac{1}{\gamma}\left(\frac{\pi}{2} + \arctan(\gamma a)\right),$$

and the corresponding integrals on the right-hand side are

$$S_\gamma = \frac{1}{2\gamma}\left(\frac{\pi}{2} + \arctan(\gamma a) + \frac{\gamma a}{1 + \gamma^2 a^2}\right)$$

and

$$T_\gamma = \frac{1}{2\gamma^3}\left(\frac{\pi}{2} + \arctan(\gamma a) - \frac{\gamma a}{1 + \gamma^2 a^2}\right).$$

Therefore,

$$\lim_{\gamma\to\infty} \frac{\left(\int f_\gamma(x)\,dx\right)^4}{\int f_\gamma^2(x)\,dx \int (x-a)^2 f_\gamma^2(x)\,dx} = 4\pi^2$$

is a lower bound for the constant in (3.18). But we have already shown that $4\pi^2$ is an upper bound, and thus this is, indeed, the sharp constant.

□

Remark 3.13 Since $\arctan t$ is strictly smaller than $\frac{\pi}{2}$ for any choice of the positive number t, we do, in fact, have *strict* inequality in (3.18) whenever f is a non-zero function such that the integrals on the right-hand side converge.

Remark 3.14 Suppose instead that $a \leq 0$. Since then, trivially, $x \leq x-a$, it follows from Carlson's inequality (3.1) that (3.18) may be complemented by the inequality

$$\left(\int_0^\infty f(x)\,dx\right)^4 \leq \pi^2 \int_0^\infty f^2(x)\,dx \int_0^\infty (x-a)^2 f^2(x)\,dx \tag{3.19}$$

in this case. In fact, as can be seen by considering the functions

$$f_\lambda(x) = \frac{\lambda}{1+\lambda^2(x-a)^2}$$

and letting $\lambda \to 0$, the constant π^2 is sharp also when $a < 0$ (cf. Section 3.9 below). Note, in particular, the remarkable jump in the sharp constant over the parameter value $a = 0$ in (3.18) and (3.19).

We will state a two-dimensional, extended variant of Theorem 3.7 in Chapter 5.

3.9 Integrals on Bounded Intervals

The results in Section 2.10 have continuous analogues. Here, we will present some recent results of L. Larsson, Z. Páles, and L.-E. Persson [52]. This gives perhaps the most straightforward way of considering inequalities of Carlson type for integrals over a bounded interval rather than on the whole real line. We cannot, however, just transfer the summation from 1 to m to integration over $(0, m)$, and expect to get a smaller constant than in the case of integration over $(0, \infty)$. Indeed, consider e.g. the case where the parameters are as in Carlson's inequality

$$\left(\int_0^m f(x)\,dx\right)^4 \leq C \int_0^m f^2(x)\,dx \int_0^m x^2 f^2(x)\,dx. \tag{3.20}$$

Let

$$f_\lambda(x) = \frac{\lambda}{\lambda^2 + x^2}.$$

Then

$$\int_0^m f_\lambda(x)\,dx = \arctan\frac{m}{\lambda},$$

$$\int_0^m f_\lambda^2(x)\,dx = \frac{1}{2\lambda}\left(\arctan\frac{m}{\lambda} + \frac{m\lambda}{\lambda^2 + m^2}\right),$$

and

$$\int_0^m x^2 f_\lambda^2(x)\,dx = \frac{\lambda}{2}\left(\arctan\frac{m}{\lambda} - \frac{m\lambda}{\lambda^2 + m^2}\right).$$

Letting $\lambda \to 0$ yields

$$\begin{aligned} C &\geq \frac{\left(\arctan\frac{m}{\lambda}\right)^4}{\frac{1}{2\lambda}\left(\arctan\frac{m}{\lambda} + \frac{m\lambda}{\lambda^2+m^2}\right)\frac{\lambda}{2}\left(\arctan\frac{m}{\lambda} - \frac{m\lambda}{\lambda^2+m^2}\right)} \\ &= \frac{4\left(\arctan\frac{m}{\lambda}\right)^4}{\left(\arctan\frac{m}{\lambda}\right)^2 - \frac{m^2\lambda^2}{(\lambda^2+m^2)^2}} \to \pi^2. \end{aligned}$$

Thus, the sharp constant is π^2 also in the inequality (3.20). In words, this can be explained by saying that there are maximizing functions for the inequality (3.1) with their mass concentrated arbitrarily close to 0 (i.e. the functions f_λ as $\lambda \to 0$). Moreover, the upper bound π^2 for the constant in (3.20) can be found as in the discrete case. We thus have the following.

Proposition 3.3 If f is a non-zero, non-negative function on $(0, m)$, $0 < m < \infty$, then

$$\left(\int_0^m f(x)\,dx \right)^4 < \pi^2 \int_0^m f^2(x)\,dx \int_0^m x^2 f^2(x)\,dx.$$

The constant π^2 is sharp.

In view of the above proposition, we alter the interval of integration slightly. Also, for this purpose, we introduce the following truncated Beta function for $0 < u < t < 1$.

$$B(\alpha, \beta; u, t) = \int_u^t (1-s)^\alpha s^\beta \frac{ds}{(1-s)s}.$$

Theorem 3.8 Suppose that $m > 1$ and $\alpha_2 > r > \alpha_1 \geq 0$. Then the inequality

$$\begin{aligned} \int_{m^{-1}}^m f(x)\,dx \leq C \left(\int_{m^{-1}}^m x^{\alpha_1} f^{r+1}(x)\,dx \right)^{\frac{\alpha_2 - r}{(\alpha_2-\alpha_1)(r+1)}} \\ \left(\int_{m^{-1}}^m x^{\alpha_2} f^{r+1}(x)\,dx \right)^{\frac{r-\alpha_1}{(\alpha_2-\alpha_1)(r+1)}} \end{aligned} \tag{3.21}$$

holds for all non-negative, measurable functions f on (m^{-1}, m). We may choose

$$C = \left[\frac{2^{\frac{1}{r}}}{\alpha_2 - \alpha_1} B\left(\frac{\alpha_2 - r}{r(\alpha_2 - \alpha_1)}, \frac{r - \alpha_1}{r(\alpha_2 - \alpha_1)}; u, t \right) \right]^{\frac{r}{r+1}}, \tag{3.22}$$

where

$$u = \frac{m^{2(\alpha_2 - r)} - 1}{m^{2(\alpha_2 - \alpha_1)} - 1} \quad \text{and} \quad t = \frac{m^{2(\alpha_2-\alpha_1)} - m^{2(r-\alpha_1)}}{m^{2(\alpha_2-\alpha_1)} - 1}.$$

Remark 3.15 As in the discrete case, the constant $C = C_{m,r,\alpha_1,\alpha_2}$ given in (3.22) tends to the sharp constant in part (a) of Theorem 3.2 with $p = q = r + 1$, $\lambda = r - \alpha_1$, and $\mu = r + \alpha_2$ (see also Theorem 4.1 in Chapter 4).

Proof of Theorem 3.8. For any $\lambda > 0$, we have by the Hölder–Rogers inequality

$$\begin{aligned}\int_{m^{-1}}^{m} f(x)\,dx &= \int_{m^{-1}}^{m} (\lambda x^{\alpha_1}+\lambda^{-1}x^{\alpha_2})^{-\frac{1}{r+1}}(\lambda x^{\alpha_1}+\lambda^{-1}x^{\alpha_2})^{\frac{1}{r+1}} f(x)\,dx \\ &\le \left(\int_{m^{-1}}^{m} \frac{dx}{(\lambda x^{\alpha_1}+\lambda^{-1}x^{\alpha_2})^{\frac{1}{r}}}\right)^{\frac{r}{r+1}} \\ &\quad \cdot \left(\int_{m^{-1}}^{m} (\lambda x^{\alpha_1}+\lambda^{-1}x^{\alpha_2}) f^{r+1}(x)\,dx\right)^{\frac{1}{r+1}} \\ &= \left(\frac{1}{\alpha_2-\alpha_1}B\right)^{\frac{r}{r+1}} \left(\lambda^{\frac{2r-(\alpha_1+\alpha_2)}{\alpha_2-\alpha_1}}(\lambda S_1+\lambda^{-1}S_2)\right)^{\frac{1}{r+1}},\end{aligned}$$

where

$$S_i = \int_{m^{-1}}^{m} x^{\alpha_i} f^{r+1}(x)\,dx, \quad i=1,2,$$

and

$$B = \int_{\frac{m^{-(\alpha_2-\alpha_1)}}{\lambda^2+m^{-(\alpha_2-\alpha_1)}}}^{\frac{m^{\alpha_2-\alpha_1}}{\lambda^2+m^{\alpha_2-\alpha_1}}} (1-s)^{\frac{\alpha_2-r}{r(\alpha_2-\alpha_1)}} s^{\frac{r-\alpha_1}{r(\alpha_2-\alpha_1)}} \frac{ds}{(1-s)s}.$$

Now, we can minimize this truncated Beta function with respect to λ, simply by differentiating. We find that the minimum is

$$B = B\left(\frac{\alpha_2-r}{r(\alpha_2-\alpha_1)}, \frac{r-\alpha_1}{r(\alpha_2-\alpha_1)}; \frac{m^{2(\alpha_2-r)}-1}{m^{2(\alpha_2-\alpha_1)}-1}, \frac{m^{2(\alpha_2-\alpha_1)}-m^{2(r-\alpha_1)}}{m^{2(\alpha_2-\alpha_1)}-1}\right).$$

We may now put

$$\lambda = \sqrt{\frac{S_2}{S_1}},$$

which yields (3.21) with C given by (3.22). □

Remark 3.16 In the next chapter, we will focus our interest on a famous result of V. I. Levin (see Theorem 4.1). This result suggests that we should look for an inequality of the type

$$\int_{m^{-1}}^{m} f(x)\,dx \le C\left(\int_{m^{-1}}^{m} x^{p-1-\lambda} f^p(x)\,dx\right)^s \left(\int_{m^{-1}}^{m} x^{q-1+\mu} f^q(x)\,dx\right)^t. \tag{3.23}$$

The proof of Theorem 3.8 cannot be directly extended to cover also this case.

Problem 6 Prove the inequality (3.23) with a constant strictly smaller than the sharp constant given by (4.3) for the corresponding parameters, preferrably the sharp constant, for each $m > 1$.

Chapter 4

Levin's Theorem

In this chapter, we will present and discuss a remarkable result of V. I. Levin [56]. In 1948, he found the following generalization of (3.1), which is, in a way, the most general inequality of its type one can hope for (see Remark 4.5 below).

Theorem 4.1 (Levin, 1948) Suppose that $p > 1$, $q > 1$, $s > 0$ and $t > 0$, and that λ and μ are any real numbers. If

$$s = \frac{\mu}{p\mu + q\lambda} \quad \text{and} \quad t = \frac{\lambda}{p\mu + q\lambda}, \tag{4.1}$$

then

$$\int_0^\infty f(x)\,dx \le C\left(\int_0^\infty x^{p-1-\lambda} f^p(x)\,dx\right)^s \left(\int_0^\infty x^{q-1+\mu} f^q(x)\,dx\right)^t \tag{4.2}$$

for all non-negative functions f, where the best constant C is given by

$$C = \left(\frac{1}{ps}\right)^s \left(\frac{1}{qt}\right)^t \left(\frac{1}{\mu+\lambda} B\Big(\frac{s}{1-s-t}, \frac{t}{1-s-t}\Big)\right)^{1-s-t}. \tag{4.3}$$

Conversely, in order for the existence of a constant C such that (4.2) holds, it is necessary that s and t are defined by (4.1).

As will be seen below, Levin's proof is a genious refinement of Bellman's proof of Theorem 3.2.

Proof. Suppose that (4.2) holds for some C. First, let $f(x) = \rho g(x)$, where $\rho > 0$ and g is some non-zero function. Then (4.2) becomes

$$\rho^{1-ps-qt} \int_0^\infty g(x)\, dx \le C \left(\int_0^\infty x^{p-1-\lambda} g(x)\, dx \right)^s \times \left(\int_0^\infty x^{q-1+\mu} g(x)\, dx \right)^t .$$

If we let $\rho \to 0$ or $\rho \to \infty$, we see that we must have

$$ps + qt = 1. \tag{4.4}$$

Moreover, by replacing x by ρx in (4.2), we get

$$\rho^{1-(p-\lambda)s-(q+\mu)t} \int_0^\infty f(x)\, dx \le C \left(\int_0^\infty x^{p-1-\lambda} f(x)\, dx \right)^s \times \left(\int_0^\infty x^{q-1+\mu} f(x)\, dx \right)^t .$$

Thus

$$(p-\lambda)s + (q+\mu)t = 1. \tag{4.5}$$

The unique solution (s, t) of the system of linear equations (4.4) and (4.5) is precisely (4.1).

Suppose now that (4.1) holds. Let

$$z = \left(\frac{k^{p-1}}{(1-k)^{q-1}} \right)^{\frac{1}{\delta}}, \tag{4.6}$$

where

$$\delta = (p-1)\mu + (q-1)\lambda.$$

Then z, regarded as a function of k, is continuous on $(0,1)$ and increases strictly from 0 to ∞ in this interval. Therefore, we can define its inverse k by the relation (4.6). Let y be any positive number. We write

$$f(x) = (1 - k(yx)) x^{-\frac{p-1-\lambda}{p}} x^{\frac{p-1-\lambda}{p}} f(x) + k(yx) x^{-\frac{q-1+\mu}{q}} x^{\frac{q-1+\mu}{q}} f(x)$$

in the integral on the left-hand side of (4.2). Thus, if P and Q are defined by

$$P^p = \int_0^\infty x^{p-1-\lambda} f^p(x)\, dx \quad \text{and} \quad Q^q = \int_0^\infty x^{q-1+\mu} f^q(x)\, dx,$$

it follows by the Hölder–Rogers inequality that

$$\begin{aligned}\int_0^\infty f(x)\,dx &\leq \left(\int_0^\infty (1-k(yx))^{p'} x^{\frac{\lambda}{p-1}} \frac{dx}{x}\right)^{\frac{1}{p'}} P \\ &\quad + \left(\int_0^\infty k(yx)^{q'} x^{-\frac{\mu}{q-1}} \frac{dx}{x}\right)^{\frac{1}{q'}} Q \\ &= \left(\int_0^\infty (1-k(z))^{p'} z^{\frac{\lambda}{p-1}} \frac{dz}{z}\right)^{\frac{1}{p'}} y^{-\frac{\lambda}{p}} P \\ &\quad + \left(\int_0^\infty k(z)^{q'} z^{-\frac{\mu}{q-1}} \frac{dz}{z}\right)^{\frac{1}{q'}} y^{\frac{\mu}{q}} Q \\ &= y^{-\frac{\lambda}{p}} UP + y^{\frac{\mu}{q}} VQ.\end{aligned}$$

Here, p' and q' are the exponents conjugate to p and q, respectively. The integrals U and V can be calculated explicitly, as follows. By the change of variables (4.6), we get

$$\frac{dz}{z} = \frac{1}{\delta}((p-1)(1-k) + (q-1)k)\frac{dk}{(1-k)k},$$

and the interval of integration transforms to $(0,1)$. Thus

$$\begin{aligned}U^{p'} &= \int_0^\infty (1-k(z))^{p'} z^{\frac{\lambda}{p-1}} \frac{dz}{z} \\ &= \frac{1}{\delta}\int_0^1 (1-k)^{p'} \left(\frac{k^{p-1}}{(1-k)^{q-1}}\right)^{\frac{\lambda}{\delta}\frac{1}{p-1}} \\ &\quad \times ((p-1)(1-k) + (q-1)k)\frac{dk}{(1-k)k} \\ &= \frac{p-1}{\delta}\int_0^1 (1-k)^{p'-\frac{\lambda}{\delta}\frac{q-1}{p-1}+1} k^{\frac{\lambda}{\delta}} \frac{dk}{(1-k)k} \\ &\quad + \frac{q-1}{\delta}\int_0^1 (1-k)^{p'-\frac{\lambda}{\delta}\frac{q-1}{p-1}} k^{\frac{\lambda}{\delta}+1} \frac{dk}{(1-k)k}\end{aligned}$$

$$
\begin{aligned}
&= \frac{p-1}{\delta} \int_0^1 (1-k)^{\frac{\mu}{\delta}+2} k^{\frac{\lambda}{\delta}} \frac{dk}{(1-k)k} \\
&+ \frac{q-1}{\delta} \int_0^1 (1-k)^{\frac{\mu}{\delta}+1} k^{\frac{\lambda}{\delta}+1} \frac{dk}{(1-k)k} \\
&= \frac{p-1}{\delta} B\Big(\frac{\mu}{\delta}+2, \frac{\lambda}{\delta}\Big) + \frac{q-1}{\delta} B\Big(\frac{\mu}{\delta}+1, \frac{\lambda}{\delta}+1\Big) \\
&= \left(\frac{p-1}{\delta} \frac{\mu+\delta}{\mu+\lambda+\delta} \frac{\mu}{\mu+\lambda} + \frac{q-1}{\delta} \frac{\lambda}{\mu+\lambda+\delta} \frac{\mu}{\mu+\lambda}\right) B\Big(\frac{\mu}{\delta}, \frac{\lambda}{\delta}\Big) \\
&= \frac{p\mu}{(p\mu+q\lambda)(\mu+\lambda)} B\Big(\frac{\mu}{\delta}, \frac{\lambda}{\delta}\Big) \\
&= \frac{ps}{\mu+\lambda} B\Big(\frac{s}{1-s-t}, \frac{t}{1-s-t}\Big).
\end{aligned}
$$

Similarly,

$$
V^{q'} = \frac{qt}{\mu+\lambda} B\Big(\frac{s}{1-s-t}, \frac{t}{1-s-t}\Big).
$$

Now, put

$$
y = \left(\frac{qt}{ps} \frac{UP}{VQ}\right)^{\frac{pq}{p\mu+q\lambda}}.
$$

Thus

$$
\begin{aligned}
\int_0^\infty f(x)\,dx &\le \left(\frac{qt}{ps}\frac{U}{V}\frac{P}{Q}\right)^{-qt} UP + \left(\frac{qt}{ps}\frac{U}{V}\frac{P}{Q}\right)^{ps} VQ \\
&= \left(\left(\frac{qt}{ps}\right)^{-qt} + \left(\frac{qt}{ps}\right)^{ps}\right) U^{ps} V^{qt} P^{ps} Q^{qt} \\
&= \left(\frac{1}{ps}\right)^{s} \left(\frac{1}{qt}\right)^{t} \left(\frac{1}{\mu+\lambda}\right)^{(p-1)s} \left(\frac{1}{\mu+\lambda}\right)^{(q-1)t} \\
&\times B\Big(\frac{s}{1-s-t}, \frac{t}{1-s-t}\Big)^{(p-1)s+(q-1)t} P^{ps} Q^{qt} \\
&= \left(\frac{1}{ps}\right)^{s} \left(\frac{1}{qt}\right)^{t} \\
&\times \left(\frac{1}{\mu+\lambda} B\Big(\frac{s}{1-s-t}, \frac{t}{1-s-t}\Big)\right)^{1-s-t} P^{ps} Q^{qt},
\end{aligned}
$$

which is (4.2) with C given by (4.3). By considering possible simultaneous cases of equality in the applications of the Hölder–Rogers inequality above, we find that equality holds in (4.2) with the constant (4.3) precisely when $f(x)$ is a multiple of

$$y^{-\frac{\mu}{q-1}} k(yx)^{\frac{1}{q-1}} x^{-\frac{\mu}{q-1}-1}$$

for some $y > 0$. Here, k is the function defined by (4.6). □

Remark 4.1 In 1955, B. Kjellberg [45] explained how to prove the inequality (4.2) with the method of calculus of variations (see also the proofs at the beginning of Chapter 3).

Remark 4.2 (Cf. [5]) We prove that Levin's inequality (4.2) is equivalent to a similar inequality, where the integrals involved are taken over the whole real line. Thus, in particular, Carlson's inequality (3.1) is equivalent to (3.6), and thus to Beurling's inequality (3.4), as was shown separately in Section 3.1 of Chapter 3.

The inequality under consideration is

$$\begin{aligned}\int_{-\infty}^{\infty} f(x)\,dx \le 2^{1-s-t} C \left(\int_{-\infty}^{\infty} |x|^{p-1-\lambda} f^p(x)\,dx\right)^s \\ \times \left(\int_{-\infty}^{\infty} |x|^{q-1+\mu} f^q(x)\,dx\right)^t,\end{aligned} \tag{4.7}$$

where C is the constant given by (4.3). If f is defined on $(0,\infty)$, extend f to an even function on $(-\infty,\infty)$ and apply (4.7). Since

$$\int_{-\infty}^{\infty} g(x)\,dx = 2\int_0^{\infty} g(x)\,dx$$

if $g(x) = f(x)$, $g(x) = f^2(x)$ or $g(x) = x^2 f^2(x)$, we get (4.2). In the other direction, we apply the elementary inequality

$$a_1^s b_1^t + a_2^s b_2^t \le 2^{1-s-t}(a_1+a_2)^s(b_1+b_2)^t, \tag{4.8}$$

valid for all non-negative numbers a_i, b_i. This can be proved as follows. Let $r = 1-s-t$. Then the Hölder–Rogers inequality can be applied with the three exponents $1/r$, $1/s$ and $1/t$:

$$a_1^s b_1^t + a_2^s b_2^t \le (1^{1/r} + 1^{1/r})^r (a_1+a_2)^s (b_1+b_2)^t =$$

$$= 2^{1-s-t}(a_1+a_2)^s(b_1+b_2)^t.$$

Let

$$a_1 = \int_{-\infty}^{0} |x|^{p-1-\lambda} f^p(x)\,dx, \quad a_2 = \int_{0}^{\infty} |x|^{p-1-\lambda} f^p(x)\,dx,$$

$$b_1 = \int_{-\infty}^{0} |x|^{q-1+\mu} f^q(x)\,dx, \quad b_2 = \int_{0}^{\infty} |x|^{q-1+\mu} f^q(x)\,dx.$$

Then, by (4.2) and (4.8)

$$\begin{aligned}
&\int_{-\infty}^{\infty} f(x)\,dx = \int_{-\infty}^{0} f(x)\,dx + \int_{0}^{\infty} f(x)\,dx \le \\
&\le C(a_1^s b_1^t + a_2^s b_2^t) \le 2^{1-s-t} C (a_1 + a_2)^s (b_1 + b_2)^t = \\
&= 2^{1-s-t} C \left(\int_{-\infty}^{\infty} |x|^{p-1-\lambda} f^p(x)\,dx \right)^s \left(\int_{-\infty}^{\infty} |x|^{q-1+\mu} f^q(x)\,dx \right)^t .
\end{aligned}$$

Remark 4.3 The inequality (4.2) is really just the first part of Bellman's Theorem 3.2. However, as stated in Theorem 4.1, Levin found the sharp constant in this inequality. By comparing the constant (4.3) with the constant (3.10) found by examining the proof of Theorem 3.2, we see that Bellman's constant is in general not sharp.

Remark 4.4 It is clear that the discrete inequality corresponding to (4.2), namely

$$\sum_{k=1}^{\infty} a_k \le C \left(\sum_{k=1}^{\infty} k^{p-1-\lambda} a_k^p \right)^s \left(\sum_{k=1}^{\infty} k^{q-1+\mu} a_k^q \right)^t , \tag{4.9}$$

with the same constant, can be proved to hold under the same conditions on s and t, by following exactly the same method as when proving the continuous version. In fact, we even have strict inequality in (4.9), unless all the a_k are zero. The special case of (4.9) where $q = p$ and $\mu = \lambda = \delta$ is just Gabriel's Theorem 2.1. Moreover, the inequalities (4.2) and (4.9) were proved by Bellman in Theorems 3.2 and 2.4, respectively, although he achieved the best constant only in the special cases $q = p$ and $\mu = \lambda$.

Remark 4.5 In Levin's inequality (4.2), put

$$f(x) = |g(x)x^d|^r \frac{1}{x}.$$

Moreover, put

$$p_0 = rp \quad \text{and} \quad p_1 = rq,$$

$$d_0 = d - \frac{\lambda}{p_0} \quad \text{and} \quad d_1 = d + \frac{\mu}{p_1}$$

and

$$\theta = \frac{p_1\lambda}{p_0\mu + p_1\lambda}.$$

The inequality then, after taking rth roots, becomes

$$\begin{aligned} &\left(\int_0^\infty |g(x)x^d|^r \frac{dx}{x}\right)^{1/r} \le C^{1/r} \\ &\times \left(\int_0^\infty |g(x)x^{d_0}|^{p_0} \frac{dx}{x}\right)^{(1-\theta)/p_0} \left(\int_0^\infty |g(x)x^{d_1}|^{p_1} \frac{dx}{x}\right)^{\theta/p_1}. \end{aligned} \tag{4.10}$$

The conditions (4.1) together with the requirements $s, t > 0$ can now be translated to

$$d = (1-\theta)d_0 + \theta d_1 \quad \text{and} \quad d_0 \neq d_1. \tag{4.11}$$

Moreover, since $p, q > 1$, we need to have $r < p_0, p_1$. We thus have the following reformulation of Levin's theorem. Recall that we use L_r^* to denote the Lebesgue spaces formed when using homogeneous measure.

Theorem 4.2 Suppose that $p_0, p_1 > r > 0$ and $0 < \theta < 1$. Suppose, moreover, that

$$w(x) = x^d, w_0(x) = x^{d_0}, w_1(x) = x^{d_1}, \quad 0 < x < \infty.$$

Then the inequality

$$\|fw\|_{L_r^*(\mathbb{R}_+)} \le C \|fw_0\|_{L_{p_0}^*(\mathbb{R}_+)}^{1-\theta} \|fw_1\|_{L_{p_1}^*(\mathbb{R}_+)}^{\theta}$$

holds for some constant C if and only if the conditions (4.11) hold.

In Chapter 5, we will give a generalization of Theorem 4.2, and thus of Theorem 4.1, to infinite cones in $\mathbb{R}^n$.

Chapter 5

Some Multi-dimensional Generalizations and Variations

5.1 Some Preliminaries

In this chapter, we will investigate some multi-dimensional generalizations of Carlson's inequality. Some guiding questions are the following.

- How should a straightforward multi-dimensional generalization of Carlson's integral inequality (3.1) look like?
- How can Levin's theorem (Theorem 4.1) be generalized to higher dimensions?

In order to introduce the results we aim to prove, we give the following two-dimensional examples.

Example 5.1 One may start by asking whether there is a constant C for which the inequality

$$\left(\iint_{\mathbb{R}_+^2} f(x,y)\,dx\,dy \right)^4 \leq C \iint_{\mathbb{R}_+^2} f^2(x,y)\,dx\,dy \iint_{\mathbb{R}_+^2} xyf^2(x,y)\,dx\,dy \tag{5.1}$$

holds for all measurable, non-negative functions f on $\mathbb{R}_+^2$. In fact, this is not the case, which can be seen by considering the characteristic functions f_R of the sets

$$\{(x,y) \in \mathbb{R}_+^2; 0 < x^2 + y^2 < R^2\}.$$

For in this case, we have

$$\iint_{\mathbb{R}_+^2} f_R(x,y)\,dx\,dy = \iint_{\mathbb{R}_+^2} f_R^2(x,y)\,dx\,dy = \frac{\pi}{4}R^2$$

and

$$\iint_{\mathbb{R}^2_+} xy f_R^2(x,y)\,dx\,dy = \frac{1}{8}R^4.$$

A lower bound for a possible constant in (5.1) is therefore

$$\frac{\pi^3}{8}R^2, \quad R > 0.$$

By letting $R \to \infty$, we see that the existence of such a constant is impossible.

Example 5.2 We consider now the possibility of finding a constant C such that the inequality

$$\left(\iint_{\mathbb{R}^2_+} f(x,y)\,dx\,dy\right)^4 \le C \iint_{\mathbb{R}^2_+} f^2(x,y)\,dx\,dy \iint_{\mathbb{R}^2_+} x^2y^2 f^2(x,y)\,dx\,dy \tag{5.2}$$

holds for all f. For $R > 1$, we let

$$f_R(x,y) = \frac{1}{1+x^2y^2}\chi_R(x,y),$$

where we denote by χ_R the characteristic function of the set

$$E_R = \{(x,y) \in \mathbb{R}^2_+; 1 < y < R\}.$$

We then have

$$\begin{aligned}\iint_{\mathbb{R}^2_+} f_R(x,y)\,dx\,dy &= \int_1^R \left(\int_0^\infty \frac{1}{1+x^2y^2}\,dx\right) dy \\ &= \int_1^R \frac{\pi}{2y}\,dy = \frac{\pi}{2}\log R,\end{aligned}$$

$$\begin{aligned}\iint_{\mathbb{R}^2_+} f_R^2(x,y)\,dx\,dy &= \int_1^R \left(\int_0^\infty \frac{1}{(1+x^2y^2)^2}\,dx\right) dy \\ &= \int_1^R \frac{\pi}{4y}\,dy = \frac{\pi}{4}\log R,\end{aligned}$$

and

$$\iint_{\mathbb{R}_+^2} x^2y^2 f_R^2(x,y)\,dx\,dy = \int_1^R \left(\int_0^\infty \frac{x^2y^2}{(1+x^2y^2)^2}\,dx \right) dy$$
$$= \int_1^R \frac{\pi}{4y}\,dy = \frac{\pi}{4}\log R.$$

Thus, any constant C in (5.2) must satisfy

$$C \geq \frac{\left(\frac{\pi}{2}\log R\right)^4}{\left(\frac{\pi}{4}\log R\right)^2} = \pi^2(\log R)^2.$$

If we let $R \to \infty$, we see that this is not possible.

Example 5.3 Our final example will concern the following inequality.

$$\left(\iint_{\mathbb{R}_+^2} f(x,y)\,dx\,dy \right)^4 \leq C \iint_{\mathbb{R}_+^2} f^2(x,y)\,dx\,dy \cdot \iint_{\mathbb{R}_+^2} (x^2+y^2)^2 f^2(x,y)\,dx\,dy. \tag{5.3}$$

Let $\lambda > 0$, and define the function $k : \mathbb{R}_+^2 \to \mathbb{R}_+$ by

$$k(x,y) = \frac{(x^2+y^2)^2}{\lambda + (x^2+y^2)^2}.$$

By the Schwarz inequality

$$\begin{aligned}\iint_{\mathbb{R}_+^2} f(x,y)\,dx\,dy &= \iint_{\mathbb{R}_+^2} (1-k(x,y))f(x,y)\,dx\,dy \\ &\quad + \iint_{\mathbb{R}_+^2} \frac{k(x,y)}{x^2+y^2}(x^2+y^2)f(x,y)\,dx\,dy \\ &\le \left(\iint_{\mathbb{R}_+^2} (1-k(x,y))^2\,dx\,dy\right)^{\frac{1}{2}} \\ &\quad \cdot \left(\iint_{\mathbb{R}_+^2} f^2(x,y)\,dx\,dy\right)^{\frac{1}{2}} \\ &\quad + \left(\iint_{\mathbb{R}_+^2} \frac{k^2(x,y)}{(x^2+y^2)^2}\,dx\,dy\right)^{\frac{1}{2}} \\ &\quad \cdot \left(\iint_{\mathbb{R}_+^2} (x^2+y^2)^2 f^2(x,y)\,dx\,dy\right)^{\frac{1}{2}}.\end{aligned}$$

As the reader may check, with the above choice of the function k, we have

$$\iint_{\mathbb{R}_+^2} (1-k(x,y))^2\,dx\,dy = \frac{\pi^2}{16}\sqrt{\lambda}$$

and

$$\iint_{\mathbb{R}_+^2} \frac{k^2(x,y)}{(x^2+y^2)^2}\,dx\,dy = \frac{\pi^2}{16}\frac{1}{\sqrt{\lambda}}.$$

Thus, if we let

$$\lambda = \frac{\iint_{\mathbb{R}_+^2} (x^2+y^2)^2 f^2(x,y)\,dx\,dy}{\iint_{\mathbb{R}_+^2} f^2(x,y)\,dx\,dy},$$

we get

$$\begin{aligned}\iint_{\mathbb{R}_+^2} f(x,y)\,dx\,dy \le \frac{\pi}{2}&\left(\iint_{\mathbb{R}_+^2} f^2(x,y)\,dx\,dy\right)^{\frac{1}{4}} \\ &\left(\iint_{\mathbb{R}_+^2} (x^2+y^2)^2 f^2(x,y)\,dx\,dy\right)^{\frac{1}{4}},\end{aligned}$$

which, after taking 4th powers, shows that (5.3) in fact holds, with

$$C = \frac{\pi^4}{16}.$$

In the next section, we prove a result (Theorem 5.1) which explains and put these examples into a more general frame.

We finally mention the following two-dimensional result by Carlson [23] (see (B.11) in Appendix B).

Proposition 5.1 (Carlson, 1935) If φ and ψ are non-negative functions, then

$$\left(\int_0^\infty \int_0^\infty \varphi(x)\psi(x)\, dx\right)^2 \le \frac{\pi^2}{2} \int_0^\infty \int_0^\infty \varphi^2(x)\psi^2(y)(x^2+y^2)\, dx\, dy.$$

We have equality when

$$\varphi(x) = \psi(x) = \frac{1}{1+x^2}.$$

This, however, only covers the case where f can be written as a product $f(x, y) = \varphi(x)\psi(y)$.

5.2 A Sharp Inequality for Cones in $\mathbb{R}^n$

Let S be a measurable subset of the unit sphere in $\mathbb{R}^n$, and define the infinite cone Ω by

$$\Omega = \left\{ x \in \mathbb{R}^n; 0 < |x| < \infty, \frac{x}{|x|} \in S \right\}. \tag{5.4}$$

Suppose that the positive, measurable functions w, w_0 and w_1, defined on Ω, are homogeneous of degrees γ, γ_0 and γ_1, respectively (we say that $v : \Omega \to \mathbb{R}_+$ is homogeneous of degree α if, for all $t > 0$ and all $x \in \Omega$ it holds that $v(tx) = t^\alpha v(x)$). Suppose that $0 < p < p_0, p_1 < \infty$, and fix $\theta \in (0, 1)$. Define

$$d = \gamma + \frac{n}{p}$$

and

$$d_i = \gamma_i + \frac{n}{p_i}, \quad i = 0, 1,$$

and define q by the relation

$$\frac{1}{q} = \frac{1}{p} - \frac{1-\theta}{p_0} - \frac{\theta}{p_1}.$$

We then have the following result by S. Barza, V. I. Burenkov, J. Pečarić and L.-E. Persson [5], which extends Levin's theorem to a multi-dimensional setting. It is very convenient to work on a cone of the type (5.4), since we can transfer a great deal of the problem to $(0, \infty)$ by switching to polar coordinates.

Theorem 5.1 Let $0 < p < p_0, p_1 < \infty$. Then the Carlson type inequality

$$\|fw\|_{L_p(\Omega,dx)} \leq C \, \|fw_0\|^{1-\theta}_{L_{p_0}(\Omega,dx)} \, \|fw_1\|^{\theta}_{L_{p_1}(\Omega,dx)} \tag{5.5}$$

holds for some constant C if and only if

$$d = (1-\theta)d_0 + \theta d_1, \tag{5.6}$$

$$d_0 \neq d_1, \tag{5.7}$$

and

$$\frac{w}{w_0^{1-\theta} w_1^{\theta}} \in L_q(S, \sigma). \tag{5.8}$$

Here, σ is used to denote surface area measure on S. In (5.5), we may use

$$\begin{aligned} C = {} & (1-\theta)^{-\frac{1-\theta}{p_0}} \theta^{-\frac{\theta}{p_1}} \left(\frac{B\left(\frac{(1-\theta)q}{p_0}, \frac{\theta q}{p_1}\right)}{p_0 p_1 |d_0 - d_1|} \right)^{\frac{1}{q}} \\ & \times \left(\frac{1}{p} - \frac{1}{q} \right)^{-\frac{1}{q}} \left\| \frac{w}{w_0^{1-\theta} w_1^{\theta}} \right\|_{L_q(S,\sigma)}, \end{aligned} \tag{5.9}$$

and this constant is sharp. Equality in (5.5) holds with the constant given by (5.9) if and only if f satisfies

$$|f(x)| = H\tilde{f}(rx) \tag{5.10}$$

for almost every x, for some $H \geq 0$, $r > 0$, where

$$\tilde{f} = \left((1-k) \frac{w^p}{w_0^{p_0}} \right)^{1/(p_0 - p)}$$

and k is defined by the implicit relation

$$\left((1-k)^{1/p_0}\frac{w}{w_0}\right)^{r_0} = \left(k^{1/p_1}\frac{w}{w_1}\right)^{r_1}, \tag{5.11}$$

where

$$\frac{1}{r_i} = \frac{1}{p} - \frac{1}{p_i}, \quad i = 0, 1.$$

Remark 5.1 By applying an interpolation technique, the inequality (5.5) can, assuming that the conditions (5.6)–(5.8) hold, be proved for all p satisfying

$$\frac{1}{p} \geq \frac{1-\theta}{p_0} + \frac{\theta}{p_1}.$$

We will, however, postpone the details until Chapter 6, where we deal with general measure spaces.

Remark 5.2 Of course, this is a strict extension of Levin's Theorem 4.1, but we can also get for example Beurling's inequality (3.6) from Theorem 5.1 by letting $S = \{-1, 1\}$ in the case $n = 1$.

Remark 5.3 The if and only if nature of Theorem 5.1 gives an explanation of why the inequality (5.1) in Example 5.1 fails. Indeed, in that case we have

$$d = 2, \quad d_0 = 1, \quad \text{and} \quad d_1 = 2.$$

Thus, there is no $\theta \in (0, 1)$ such that (5.6) holds.

Remark 5.4 In Example 5.2, we replaced xy by x^2y^2 in the last integral on the right-hand side. Thus, in this case, we have

$$w(x,y) = 1, \quad w_0(x,y) = 1, \quad \text{and} \quad w_1(x,y) = xy,$$

so that

$$d = 2, \quad d_0 = 1, \quad \text{and} \quad d_1 = 3.$$

Thus (5.6) holds with $\theta = \frac{1}{2}$, and we clearly also have (5.7). However, since

$$\frac{1}{q} = \frac{1}{1} - \frac{\frac{1}{2}}{2} - \frac{\frac{1}{2}}{2} = \frac{1}{2},$$

or $q = 2$, and therefore

$$\left(\frac{w(x,y)}{w_0^{1-\theta}(x,y)w_1^{\theta}(x,y)}\right)^q = \frac{1}{xy}.$$

Therefore, the condition (5.8) translates to

$$\int_0^{\frac{\pi}{2}} \frac{1}{\cos\varphi \sin\varphi}\, d\varphi < \infty,$$

which is clearly not the case. Thus, in view of Theorem 5.1, it is impossible to find a constant C such that (5.2) holds.

Remark 5.5 In Example 5.1 above, we used

$$w(x,y) = 1, w_0(x,y) = 1, \quad \text{and} \quad w_1(x,y) = x^2 + y^2.$$

With $d = 2$, $d_0 = 1$, $d_1 = 3$, and $\theta = \frac{1}{2}$, we see that (5.6) and (5.7) hold. Moreover, the quotient

$$\frac{w}{w_0^{1-\theta} w_1^{\theta}}$$

is constant on the part of the unit circle contained in $\mathbb{R}^2_+$. Thus (5.8) holds as well. We leave it to the reader to find out whether the constant

$$\frac{\pi^4}{16}$$

found in Example 5.3 is sharp.

Proof of Theorem 5.1. We note first that, when switching to polar coordinates $x = \rho\sigma$, where

$$\rho = |x| \quad \text{and} \quad \sigma = \frac{x}{|x|},$$

we get

$$\begin{aligned}
\|fw\|^p_{L_p(\Omega,dx)} &= \int_\Omega |f(x)w(x)|^p\, dx \\
&= \int_S \int_0^\infty |f(\rho\sigma)w(\rho\sigma)|^p \rho^{n-1}\, d\rho d\sigma \\
&= \int_S w^p(\sigma) \int_0^\infty |f(\rho\sigma)|^p \rho^{p\gamma+n-1}\, d\rho d\sigma \\
&= \int_S w^p(\sigma) \int_0^\infty |f(\rho\sigma)\rho^d|^p \frac{d\rho}{\rho} d\sigma,
\end{aligned}$$

and similarly for the other norms, and hence we can apply Theorem 4.2 from the previous chapter on the inner integral. More precisely, if the conditions (5.6)–(5.8) hold, then for any σ

$$\int_0^\infty |f(\rho\sigma)\rho^d|^p \frac{d\rho}{\rho} \leq C^p \left(\int_0^\infty |f(\rho\sigma)\rho^{d_0}|^{p_0} \frac{d\rho}{\rho}\right)^{(1-\theta)p/p_0} \times \left(\int_0^\infty |f(\rho\sigma)\rho^{d_1}|^{p_1} \frac{d\rho}{\rho}\right)^{\theta p/p_1},$$

and hence by the Hölder–Rogers inequality on the measure space (S,σ) with the three parameters

$$\frac{q}{p}, \frac{p_0}{(1-\theta)p} \quad \text{and} \quad \frac{p_1}{\theta p},$$

it follows that

$$\begin{aligned}
\|fw\|^p_{L_p(\Omega,dx)} &\leq C^p \int_S w^p(\sigma) \left(\int_0^\infty |f(\rho\sigma)\rho^{d_0}|^{p_0} \frac{d\rho}{\rho}\right)^{(1-\theta)p/p_0} \\
&\quad \cdot \left(\int_0^\infty |f(\rho\sigma)\rho^{d_1}|^{p_1} \frac{d\rho}{\rho}\right)^{\theta p/p_1} d\sigma \\
&= C^p \int_S \left(\frac{w(\sigma)}{w_0^{1-\theta}(\sigma)w_1^\theta(\sigma)}\right)^p \\
&\quad \cdot \left(\int_0^\infty |f(\rho\sigma)w_0(\rho\sigma)|^{p_0}\rho^{n-1}\,d\rho\right)^{(1-\theta)p/p_0} \\
&\quad \cdot \left(\int_0^\infty |f(\rho\sigma)w_1(\rho\sigma)|^{p_1}\rho^{n-1}\,d\rho\right)^{\theta p/p_1} d\sigma \\
&\leq C^p \left(\int_S \left(\frac{w(\sigma)}{w_0^{1-\theta}(\sigma)w_1^\theta(\sigma)}\right)^q d\sigma\right)^{p/q} \\
&\quad \cdot \left(\int_S \int_0^\infty |f(\rho\sigma)w_0(\rho\sigma)|^{p_0}\rho^{n-1}\,d\rho d\sigma\right)^{(1-\theta)p/p_0} \\
&\quad \cdot \left(\int_S \int_0^\infty |f(\rho\sigma)w_1(\rho\sigma)|^{p_1}\rho^{n-1}\,d\rho d\sigma\right)^{\theta p/p_1} \\
&= C^p \left\|\frac{w}{w_0^{1-\theta}w_1^\theta}\right\|^p_{L_q(S,d\sigma)} \|fw_0\|^{(1-\theta)p}_{L_{p_0}(\Omega,dx)} \|fw_1\|^{\theta p}_{L_{p_1}(\Omega,dx)}.
\end{aligned}$$

Thus the conditions are sufficient. Suppose that f is defined by (5.10). It can be shown, by using the definition of the function k together with the homogeneity of the weights w_*, that for such f, we have exactly

$$\frac{\|fw\|_p}{\|fw_0\|_{p_0}^{1-\theta}\|fw_1\|_{p_1}^{\theta}} = C,$$

where C is given by (5.9).

Suppose now that (5.5) holds for some constant C. Theorem 4.2 then implies that the conditions (5.6) and (5.7) hold. We need to show that also (5.8) is true. We may assume that $\tau = d_0 - d_1 > 0$. Define the function $h : \mathbb{R}_+ \to (0,1)$ by the implicit relation

$$z^\rho = \frac{h(z)^{p/(p_1-p)}}{(1-h(z))^{p/(p_0-p)}}, \tag{5.12}$$

where

$$\rho = \frac{\tau}{q}\frac{pp_0}{p_0-p}\frac{pp_1}{p_1-p},$$

and define ω on S by

$$\omega = \left(\left(\frac{w}{w_0}\right)^{1/r_1}\left(\frac{w_1}{w}\right)^{1/r_0}\right)^{q/\tau p}.$$

Moreover, let

$$f_0(y) = (1-h(\omega(\sigma)x))^{r_0/pr_0'}\frac{w(y)^{r_0/r_0'}}{w_0(y)^{r_0}},$$

where $y = \sigma x \in \Omega$. By (5.12) it follows that

$$\begin{aligned}
\int_\Omega |f_0(y)w(y)|^p\frac{dy}{|y|^n} &= \int_S\int_0^\infty (1-h(\omega(\sigma)x))^{r_0'/r_0}\left(\frac{w(\sigma x)}{w_0(\sigma x)}\right)^{r_0p}\frac{dx}{x}\,d\sigma \\
&= \int_S\left(\frac{w}{w_0}\right)^{r_0p}\int_0^\infty h(\omega(\sigma)x)^{r_1/r_1'}(\omega(\sigma)x)^\rho x^{(d-d_0)r_0p}\frac{dx}{x}\,d\sigma \\
&= \int_S\left(\frac{w}{w_0}\right)^{r_0p}\int_0^\infty h(\omega(\sigma)x)^{r_1/r_1'}(\omega(\sigma)x)^\rho x^{-\theta\tau r_0p}\frac{dx}{x}\,d\sigma.
\end{aligned}$$

In the x-integral, we make the substitution $t = \omega(\sigma)x$, which yields

$$\begin{aligned}
\int_S\left(\frac{w}{w_0}\right)^{r_0p}\omega^{-\theta\tau r_0p}d\sigma\int_0^\infty h(t)^{r_1/r_1'}t^{\rho-\theta\tau r_0p} \\
= (C')^p\int_S\left(\frac{w}{w_0^{1-\theta}w_1^\theta}\right)^q d\sigma,
\end{aligned}$$

where the constant C' may be calculated explicitly in terms of the Beta function. In a similar manner, we obtain

$$\int_{\Omega} |f_0(y)w_i(y)|^{p_i} \frac{dy}{|y|^n} = C_i^{p_i} \int_S \left(\frac{w}{w_0^{1-\theta} w_1^{\theta}} \right)^q d\sigma.$$

It follows that if (5.5) holds, then

$$\begin{aligned} &C' \left(\int_S \left(\frac{w}{w_0^{1-\theta} w_1^{\theta}} \right)^q d\sigma \right)^{\frac{1}{p}} \\ &\leq C C_0^{1-\theta} \left(\int_S \left(\frac{w}{w_0^{1-\theta} w_1^{\theta}} \right)^q d\sigma \right)^{\frac{1-\theta}{p_0}} \\ &\quad C_1^{\theta} \left(\int_S \left(\frac{w}{w_0^{1-\theta} w_1^{\theta}} \right)^q d\sigma \right)^{\frac{\theta}{p_1}} \end{aligned}$$

or

$$\left\| \frac{w}{w_0^{1-\theta} w_1^{\theta}} \right\|_{L_q(S)} \leq \frac{C C_0^{1-\theta} C_1^{\theta}}{C'}.$$

The proof is complete. □

5.3 Some Variations on the Multi-dimensional Theme

Various multi-dimensional versions of inequalities of Carlson type have been proved by several authors over the years. We collect some of the results below.

5.3.1 *Kjellberg Revisited*

We recall that the inequality (3.6) implies that if two integrals of the form

$$\int_{-\infty}^{\infty} x^{2\alpha} |f(x)|^2 \, dx$$

are finite (in the present case with $\alpha = 0$ and $\alpha = 1$, respectively), then so is the integral of $|f|$. This way of looking at Carlson's inequality was generalized in 1942 by B. Kjellberg [43], who studied the conditions under which the finiteness of a number of *moments* M_j, defined by

$$M_j^2 = \int_{-\infty}^{\infty} \cdots \int_{-\infty}^{\infty} |f(x_1, \ldots, x_n)|^2 x_1^{2\alpha_{j1}} \cdots x_n^{2\alpha_{jn}} \, dx_1 \cdots dx_n,$$

$$j = 1, \ldots, m,$$

implies the integrability of $|f|$ on $\mathbb{R}^n$. The main theorem is the following.

Theorem 5.2 (Kjellberg, 1942) A necessary and sufficient condition for the finiteness of the moments M_j, $j = 1, \ldots, m$, to imply the integrability of f on $\mathbb{R}^n$ is that

$$\left(\frac{1}{2}, \ldots, \frac{1}{2}\right) \in \mathbb{R}^n$$

is an interior point of the convex hull of the points

$$(\alpha^{j1}, \ldots, \alpha^{jn}), \quad j = 1, \ldots, m.$$

Suppose that $S = \{Q_1, \ldots, Q_m\}$, where each Q_j is a point $(\alpha_{j1}, \ldots, \alpha_{jn})$ giving rise to a moment M_j. Let $P(x) = \sum_{j=1}^{m} x^{2Q_j}$, where standard multi-index notation is used to define x^{2Q}. By integrating the inequality $(ab)^{1/2} \leq \frac{1}{2}(a+b)$ applied to $a = P^{-1}$ and $b = |f|^2 P$, Kjellberg noted that

$$\int_{\mathbb{R}^n} |f|\, dx \leq \frac{1}{2}\left(\int_{\mathbb{R}^n} \frac{dx}{P} + \sum_{j=1}^{m} M_j^2\right).$$

Thus, in order for the finiteness of the moments to imply that of $\int_{\mathbb{R}^n} |f|\, dx$, it is sufficient that P^{-1} is integrable. If P^{-1} is not integrable, however, there is f for which $\sum_j M_j^2 = 1$ but $\int |f| = \infty$. In one dimension, P^{-1} is integrable precisely when P contains two terms x^{2a} and x^{2b}, where $a < \frac{1}{2} < b$. Kjellberg's result thus says that this intuitive description is true also in higher dimensions.

We mention that Kjellberg's theorem is, apart from Proposition 5.1, the first multi-dimensional result of its kind.

5.3.2 *Andrianov*

In 1967, F. I. Andrianov [2] gave a proof of a quite straightforward generalization of Levin's theorem, based on Hardy's method. Suppose that

$Q = Q(x_1, \ldots, x_n)$ is homogeneous of degree n. Let

$$\begin{aligned}
x_1 &= r\cos\varphi_1, \\
x_2 &= r\sin\varphi_1\cos\varphi_2, \\
&\vdots \\
x_{n-1} &= r\sin\varphi_1\sin\varphi_2\cdots\sin\varphi_{n-2}\cos\varphi_{n-1}, \\
x_n &= r\sin\varphi_1\sin\varphi_2\cdots\sin\varphi_{n-2}\sin\varphi_{n-1}
\end{aligned}$$

and put

$$\tilde{Q}(\varphi_1, \ldots, \varphi_{n-1}) = \frac{Q}{r^n}.$$

We assume that

$$I = \int_0^{\pi/2} \cdots \int_0^{\pi/2} \frac{\sin^{n-2}\varphi_1 \sin^{n-3}\varphi_2 \cdots \sin\varphi_{n-2} d\varphi_1 d\varphi_2 \cdots d\varphi_{n-1}}{\tilde{Q}(\varphi_1, \ldots, \varphi_{n-1})} \tag{5.13}$$

is finite. We let $x = (x_1, \ldots, x_n)$, and write $x \geq 0$ for $x_j \geq 0$, $j = 1, \ldots, n$.

Theorem 5.3 (Andrianov, 1967) Suppose that $f(x) \geq 0$, $x \geq 0$, and $p, q, \lambda, \mu > 0$. Then

$$\int_{\mathbb{R}^n_+} f(x)\,dx \leq C\left(\int_{\mathbb{R}^n_+} Q^{-1-\lambda}(x) f^p(x)\,dx\right)^s \left(\int_{\mathbb{R}^n_+} Q^{q-1+\mu}(x) f^q(x)\,dx\right)^t,$$

where

$$s = \frac{\mu}{p\mu + q\lambda}, \quad t = \frac{\lambda}{p\mu + q\lambda}$$

and

$$C = \left(\frac{1}{ps}\right)^s \left(\frac{1}{qt}\right)^t \left(\frac{I}{n}\frac{1}{\lambda+\mu} B\left(\frac{s}{1-s-t}, \frac{t}{1-s-t}\right)\right)^{1-s-t}$$

($B(\cdot,\cdot)$ denotes the Beta function). The constant C is sharp.

Remark 5.6 Of course, we need to make sure that the arguments to the Beta function above are positive. This is the case if $p, q > 1$.

Remark 5.7 Note that the constant C depends on the integral I defined by (5.13), which is merely assumed to be finite and might be difficult to find the numerical value of. However, Andrianov gave an explicit expression for the best constant in the special case where Q is the nth power of a homogeneous polynomial of degree 1.

Remark 5.8 Theorem 5.3 is really a special case of Theorem 5.1. Indeed, $\mathbb{R}_+^n$ can be written as a cone Ω as in (5.4) if we choose S to be the subset of the unit sphere consisting of the points all of whose coordinates are positive.

Andrianov also proved the following result, with more than two factors appearing on the right-hand side, which is a continuous, multi-dimensional version of Theorem 2.2.

Theorem 5.4 (Andrianov, 1967) Suppose that $f(x) \geq 0$, $x \geq 0$, and $q > 0$, k is a positive integer. Then

$$\left(\int_{x\geq 0} f(x)\,dx\right)^{(k+1)(2k+1)} \leq C \prod_{j=0}^{2k} \int_{x\geq 0} Q^{k+(j-k)q}(x) f^{k+1}(x)\,dx,$$

where the constant

$$C = (2k+1)^{2k+1} \left(\frac{I}{qn}\right)^{k(2k+1)} \prod_{j=0}^{2k} \binom{2k}{j}$$

is sharp.

Remark 5.9 Here, as well, the constant depends on the integral I. In [2], this was calculated in the special case where Q is a linear combination of monomials of degree n.

5.3.3 *Pigolkin*

G. M. Pigolkin [74] also used the Hardy method to prove his multi-dimensional version of the inequality. Here, we consider functions $f = f(x_1, \ldots, x_{n-1})$ of $n-1$ variables and a non-degenerate $(n-1) \times (n-1)$ matrix (μ_{ij}) with determinant

$$G_\mu = |\mu_{ij}|.$$

Theorem 5.5 (Pigolkin, 1970) Suppose that the inverse matrix $(\mu_{ij})^{-1} = (\nu_{ij})$ satisfies

$$\sigma_j = \sum_{i=1}^{n-1} \nu_{ij} > 0.$$

If $0 < \gamma < 1$, then

$$\left(\int f\,dx\right)^{\frac{\sigma+\gamma}{\sigma(1-\gamma)}} \le C \int f^{\frac{\sigma+\gamma}{\sigma}}\,dx \prod_{i=1}^{n-1}\left(\int f^{\frac{\sigma+\gamma}{\sigma}} \prod_{j=1}^{n-1} |x_j|^{\mu_{ij}}\,dx\right)^{\nu_i}. \quad (5.14)$$

If

$$\delta = \frac{1}{1-\gamma},$$
$$\nu = \prod_{i=1}^{n-1} \nu_i^{\nu_i},$$
$$\nu_i = \frac{\gamma}{1-\gamma}\frac{\sigma_i}{\sigma},$$
$$\sigma = \sum_{j=1}^{n-1} \sigma_j,$$

then

$$C = 2^{\frac{(n-1)\gamma}{\sigma(1-\gamma)}} \frac{\delta^\delta}{\nu} \left(\frac{\Gamma\left(\frac{1-\gamma}{\gamma}\sigma\right)}{|G_\mu|\Gamma\left(\frac{\sigma}{\gamma}\right)} \prod_{i=1}^{n-1}\Gamma(\sigma_i)\right)^{\frac{\gamma}{\sigma(1-\gamma)}}$$

is the sharp constant, and equality holds in (5.14) with this constant if and only if f has the form

$$f(x) = \alpha_0 \left(\sum_{i=1}^{n-1}\left(\alpha_i + \frac{1}{\alpha_i}\prod_{j=1}^{n-1}|x_j|^{\mu_{ij}}\right)\right)^{-\frac{\sigma}{\gamma}}$$

for some $\alpha_0 \ge 0$, $\alpha_i > 0$, $i = 1, \ldots, n-1$.

5.3.4 *Bertolo–Fernandez*

By using an induction argument, J. I. Bertolo and D. L. Fernandez [14] proved a multi-dimensional version involving mixed L_P-norms, which was used in connection with the multi-dimensional Mellin transform (see [13]). We recover the main result below.

Suppose that $P = (p_1, \ldots, p_n)$, where $1 \le p_j < \infty$, $j = 1, \ldots, n$. The space L_P^* is determined by the finiteness of norms, defined inductively as follows. If $n = 1$, we put

$$\|f\|_{L^*_P(\mathbb{R}_+)} = \left(\int_0^\infty |f(x_1)|^{p_1} \frac{dx_1}{x_1} \right)^{1/p_1}.$$

If $n > 1$, and $L^*_{\hat{P}}(\mathbb{R}^{n-1}_+)$ has been defined with $\hat{P} = (p_1, \ldots, p_{n-1})$, let $P = (\hat{P}, p_n)$. Then $L^*_P(\mathbb{R}^n_+)$ is defined via the norm

$$\|f\|_{L^*_P(\mathbb{R}^n_+)} = \left(\int_0^\infty \|f(\cdot, x_n)\|^{p_n}_{L^*_{\hat{P}}(\mathbb{R}^{n-1}_+)} \frac{dx_n}{x_n} \right)^{1/p_n}.$$

Moreover, if $k = (k_1, \ldots, k_n)$, we define

$$\Theta(k) = \prod_{j=1}^n \theta(k_j),$$

where

$$\theta(k_j) = 1 - k_j + (-1)^{k_j+1}.$$

Theorem 5.6 (Bertolo–Fernandez, 1984) Let $\Theta = (\theta_1, \ldots, \theta_n)$, where $0 < \theta_j < 1$, and let $P_0 = (p_0^1, \ldots, p_0^n)$ and $P_1 = (p_1^1, \ldots, p_1^n)$, where $1 \leq p_i^j < \infty$, $i = 0, 1$, $j = 1, \ldots, n$. If $k = (k_1, \ldots, k_n)$ is such that each k_j is either 0 or 1, put $P_k = (p_{k_1}^1, \ldots, p_{k_n}^n)$. Then

$$\|f\|_{L^*_1(\mathbb{R}^n_+)} \leq C \prod_k \|x^{k-\Theta} f\|^{\Theta(k)}_{L^*_{p_k}(\mathbb{R}^n_+)} \tag{5.15}$$

for all measurable functions f on $\mathbb{R}^n_+$. The product on the right-hand side of (5.15) is taken over all possible k with elements 0 and 1. We may choose

$$C = \prod_{j=1}^n C(p_0^j, p_1^j, \theta_j),$$

where each $C(p_0^j, p_1^j, \theta_j)$ is any constant for which the corresponding one-dimensional inequality holds (see e.g. Theorem 4.1).

5.3.5 *Barza et al.*

In 1998, S. Barza, J. Pečarić, and L.-E. Persson [6] also discussed the possibility to prove some multi-dimensional version of Theorem 3.7 (with weights involving factors of the type $(x - a)^\alpha$). In particular, they obtained the following result.

Theorem 5.7 Suppose that f is a measurable function on $\mathbb{R}^2_+$, and let a_1, a_2, a_3 be any real numbers, and $\gamma_1, \gamma_2, \gamma_3 > 1$. Then

$$\iint_{\mathbb{R}^2_+} |f(x,y)|\,dx\,dy \le C \left(\iint_{\mathbb{R}^2_+} f^2(x,y)\,dx\,dy \right)^{\frac{1}{2\gamma_3'\gamma_1'}} \left(\iint_{\mathbb{R}^2_+} |x-a_1|^{\gamma_1} f^2(x,y)\,dx\,dy \right)^{\frac{1}{2\gamma_3'\gamma_1}} \left(\iint_{\mathbb{R}^2_+} |y-a_3|^{\gamma_3} f^2(x,y)\,dx\,dy \right)^{\frac{1}{2\gamma_3\gamma_2'}} \left(\iint_{\mathbb{R}^2_+} |x-a_2|^{\gamma_2} |y-a_3|^{\gamma_3} f^2(x,y)\,dx\,dy \right)^{\frac{1}{2\gamma_2\gamma_3}}.$$

We may choose

$$C = C_{\gamma_1,a_1}^{1/\gamma_3'} C_{\gamma_2,a_2}^{1/\gamma_3} C_{\gamma_3,a_3},$$

where

$$C^2_{\gamma_i,a_i} = 2^{m_i}(\gamma_i - 1)^{-\frac{1}{\gamma_i'}} \frac{\pi}{\sin(\pi/\gamma_i)},$$

with $m_i = 0$ if $a_i \le 0$ and $m_i = 1$ if $a_i > 0$, $i = 1, 2, 3$.

5.3.6 *Kamaly*

In 2000, A. Kamaly [40] (see also [38]) proved a discrete, multi-dimensional version of Beurling's inequality, which he used to get a sharp, local version of the Hausdorff–Young inequality (see Chapter 9). For a related result, see also G. W. Hedstrom [35].

Let $\mathbb{T}^n = \mathbb{R}^n/\mathbb{Z}^n$ denote the n-dimensional torus, and define the kth Fourier coefficient $\hat{f}(k)$ of an integrable function $f : \mathbb{T}^n \to \mathbb{C}$ by

$$\hat{f}(k) = \int_{\mathbb{T}^n} f(x) e^{-2\pi i k\cdot x}\,dx, \quad k \in \mathbb{Z}^n.$$

$A(\mathbb{T}^n)$ is, by definition, the vector space of integrable functions having absolutely convergent Fourier series. We equip $A(\mathbb{T}^n)$ with the norm

$$\|f\|_{A(\mathbb{T}^n)} = \sum_{k\in\mathbb{Z}^n} |\hat{f}(k)|,$$

under which it becomes a Banach space. Note that the discrete version

$$\sum_{k\in\mathbb{Z}} |\hat{f}| \leq \sqrt{2\pi} \left(\int_0^{2\pi} |f(t)|^2 \, dt \int_0^{2\pi} |f'(t)|^2 \, dt \right)^{\frac{1}{4}}$$

of Beurling's inequality (3.4) can, with a suitable normalization of the Fourier transform, be written as

$$\|\hat{f}\|_{A(\mathbb{T})} \leq C \, \|f\|_{L_2(\mathbb{T})}^{\frac{1}{2}} \, \|f'\|_{L_2(\mathbb{T})}^{\frac{1}{2}} \, .$$

In what follows, we present Kamaly's generalization of this inequality.

If $\gamma = (\gamma_1, \ldots, \gamma_n) \in \mathbb{Z}_+^n$ is a multi-index, then the differential operator D^γ is defined by

$$D^\gamma f = \frac{\partial^{|\gamma|} f}{\partial x_1^{\gamma_1} \cdots \partial x_n^{\gamma_n}},$$

where $|\gamma| = \gamma_1 + \ldots + \gamma_n$.

Theorem 5.8 (Kamaly, 2000) Let $f \in A(\mathbb{T}^n)$. Suppose that $1 < r \leq 2$ and that the positive integer α is such that $\alpha > \frac{n}{r}$. If $\hat{f}(0) = 0$, then

$$\|f\|_{A(\mathbb{T}^n)} \leq C \|f\|_r^{1-\frac{n}{r\alpha}} \left(\sum_{|\gamma|=\alpha} \|D^\gamma f\|_r \right)^{\frac{n}{r\alpha}} .$$

In the general case, we get

$$\|f\|_{A(\mathbb{T}^n)} \leq \|f\|_1 + C \|f\|_r^{1-\frac{n}{r\alpha}} \left(\sum_{|\gamma|=\alpha} \|D^\gamma f\|_r \right)^{\frac{n}{r\alpha}} .$$

In both cases, the constant C depends only on α, n, and r.

5.4 Some Further Generalizations

In this section, we will point out the fact that, by using some ideas presented in this book, we can further generalize some of the results from the previous chapters in a multi-dimensional setting.

5.4.1 *A Multi-dimensional Extension of Theorem 3.6*

We prove the following multi-dimensional extension of Theorem 3.6 (see [49]). In particular, we obtain the Yang–Fang result (see [84]), except that we do not give any estimates of the constant, upon letting $n = 1$, $p = q$, $r = s$ and $\gamma = 1$ if we consider functions f supported on $\mathbb{R}_+$.

Theorem 5.9 Let n be a positive integer. Suppose that $p, q > 2$, $0 < \alpha < n$, and $r, s \in \mathbb{R}$. Suppose, moreover, that for some positive constants M and γ, the function $g : \mathbb{R}^n \to (0, \infty)$ satisfies

$$g(x) \geq M|x|^\gamma. \tag{5.16}$$

Then there is a constant C, independent of M, α and γ, such that

$$\begin{aligned} \left(\int_{\mathbb{R}^n} |f(x)|\, dx \right)^{p+q} &\leq \frac{C}{\alpha^2 M^{2n/\gamma}} \\ &\times \int_{\mathbb{R}^n} g(x)^{(n-\alpha)/\gamma} |f(x)|^{p(1+2r-rp)}\, dx \int_{\mathbb{R}^n} g(x)^{(n+\alpha)/\gamma} |f(x)|^{q(1+2s-sq)}\, dx \\ &\times \left(\int_{\mathbb{R}^n} |f(x)|^{rp}\, dx \right)^{p-2} \left(\int_{\mathbb{R}^n} |f(x)|^{sq}\, dx \right)^{q-2}. \end{aligned} \tag{5.17}$$

We give the following simple example to illustrate the usefulness of the weaker assumptions on the function g in Theorem 5.9 in comparison with those of Theorem 3.6.

Example 5.4 Let a and b be positive real numbers, and define $g : [0, \infty) \to (0, \infty)$ by

$$g(x) = \begin{cases} a, & 0 \leq x < b, \\ 1 + a\sqrt{\frac{x}{b}}, & b \leq x < \infty. \end{cases}$$

Then g is not differentiable (not even continuous at the point $x = b$) and $g(0) = a \neq 0$. Thus two of the conditions on g in the hypotheses of Theorem 3.6 fail. Nevertheless, g is an admissible function for Theorem 5.9, since (5.16) holds with $\gamma = \frac{1}{2}$ and $M = \frac{a}{\sqrt{b}}$.

Remark 5.10 Although our proof shows that we can get away with somewhat weaker assumptions on the function g than those imposed in Theorem 3.6, the condition $\lim_{x\to\infty} g(x) = \infty$ cannot be relaxed too much. More precisely, we cannot let the function g be essentially bounded. To see this, we consider the one-dimensional case, and assume that $0 \le g(x) \le G$ almost everywhere on $[0, \infty)$. For $R > 0$, let f_R be the characteristic function of the interval $[0, R)$. Then for any t

$$\int_0^\infty |f_R(x)|^t \, dx = R,$$

and

$$\int_0^\infty g(x)^{(1\mp\alpha)/\gamma} |f_R(x)|^t \, dx \le R G^{(1\mp\alpha)/\gamma}.$$

Hence

$$\frac{\left(\int |f_R| \, dx\right)^{p+q} \left(\int |f_R|^{rp} \, dx\right)^{-(p-2)} \left(\int |f_R|^{sq} \, dx\right)^{-(q-2)}}{\int g^{(1-\alpha)/\gamma} |f_R|^{p(1+2r-rp)} \, dx \int g^{(1+\alpha)/\gamma} |f_R|^{q(1+2s-sq)} \, dx}$$
$$\ge \frac{R^{p+q}}{R^{p-2} R^{q-2} R G^{(1-\alpha)/\gamma} R G^{(1+\alpha)/\gamma}} = \frac{R^2}{G^{2/\gamma}},$$

which tends to infinity as $R \to \infty$, and thus the inequality cannot hold for any finite constant. It is clear that this also shows failure of (5.17) in each multi-dimensional case.

To prove Theorem 5.9, we apply Theorem 6.1 of Chapter 6, which we quote below in a special case, suitable for our present needs.

Lemma 5.1 Let (Ω, μ) be a measure space on which weights $w \ge 0$, $w_0 > 0$ and $w_1 > 0$ are defined. Suppose that $p_0, p_1 > 1$ and $0 < \theta < 1$. Suppose also that there is a constant A such that

$$\mu\left(\left\{\omega; 2^m \le \frac{w_0(\omega)}{w_1(\omega)} < 2^{m+1}\right\}\right) \le A, \quad m \in \mathbb{Z} \tag{5.18}$$

and that

$$\frac{w}{w_0^{1-\theta} w_1^{\theta}} \in L_\infty(\Omega, \mu). \tag{5.19}$$

Then there is a constant C such that

$$\|fw\|_{L_1} \le C \|fw_0\|_{L_{p_0}}^{1-\theta} \|fw_1\|_{L_{p_1}}^{\theta}. \tag{5.20}$$

The constant C can be chosen to have the form

$$C = C_0 A^{1-(1-\theta)/p_0 - \theta/p_1},$$

where C_0 does not depend on A.

Proof of Theorem 5.9. Let $\Omega = \mathbb{R}^n$ and define the measure μ on Ω by

$$d\mu(x) = \frac{dx}{|x|^n},$$

where dx denotes Lebesgue measure in $\mathbb{R}^n$. Define, on Ω, $w(x) = |x|^n$, $w_0(x) = |x|^{n-\alpha/p}$ and $w_1(x) = |x|^{n+\alpha/q}$, and let $p_0 = p'$, $p_1 = q'$ and

$$\theta = \frac{q}{p+q}.$$

Then

$$\frac{w}{w_0^{1-\theta} w_1^{\theta}} \equiv 1 \in L_\infty,$$

so (5.19) is satisfied. Moreover,

$$\frac{w_0(x)}{w_1(x)} = x^{-\alpha(\frac{1}{p}+\frac{1}{q})}.$$

Let

$$\tau = \alpha \left(\frac{1}{p} + \frac{1}{q} \right) > 0.$$

Thus $w_0(x)/w_1(x) \in [2^m, 2^{m+1})$ if and only if

$$2^{-(m+1)/\tau} < |x| \leq 2^{-m/\tau}.$$

Hence, using polar coordinates, we get

$$\mu\left(\left\{ \frac{w_0}{w_1} \in [2^m, 2^{m+1}) \right\} \right) = \omega_n \int_{2^{-(m+1)/\tau}}^{2^{-m/\tau}} \frac{dr}{r} = \frac{\omega_n \log 2}{\tau},$$

where ω_n denotes the surface area of the unit sphere in $\mathbb{R}^n$, which shows that (5.18) holds with

$$A = \frac{\omega_n \log 2}{\tau} = \frac{A_0}{\alpha}.$$

Thus (5.20) implies

$$\begin{aligned}\int_{\mathbb{R}^n} |f(x)|\,dx &= \int_\Omega |f(x)w(x)|\,d\mu(x)\\ &\le C\left(\int_\Omega |f(x)w_0(x)|^{p_0}\,d\mu(x)\right)^{(1-\theta)/p_0}\\ &\times \left(\int_\Omega |f(x)w_1(x)|^{p_1}\,d\mu(x)\right)^{\theta/p_1}\\ &= C\left(\int_{\mathbb{R}^n} |x|^{(n-\alpha)/(p-1)}|f(x)|^{p'}\,dx\right)^{(p-1)/(p+q)}\\ &\times \left(\int_{\mathbb{R}^n} |x|^{(n+\alpha)/(q-1)}|f(x)|^{q'}\,dx\right)^{(q-1)/(p+q)}\end{aligned}$$

or

$$\begin{aligned}\left(\int_{\mathbb{R}^n} |f(x)|\,dx\right)^{p+q} &\le C^{p+q}\\ &\times \left(\int_{\mathbb{R}^n} |x|^{(n-\alpha)/(p-1)}|f(x)|^{p'}\,dx\right)^{p-1}\\ &\times \left(\int_{\mathbb{R}^n} |x|^{(n+\alpha)/(q-1)}|f(x)|^{q'}\,dx\right)^{q-1}.\end{aligned}$$

We write

$$|x|^{(n-\alpha)/(p-1)}|f(x)|^{p'} = \left(|x|^{(n-\alpha)/(p-1)}|f(x)|^{p'(1+2r-rp)}\right)|f(x)|^{p'r(p-2)}$$

and

$$|x|^{(n+\alpha)/(q-1)}|f(x)|^{q'} = \left(|x|^{(n+\alpha)/(q-1)}|f(x)|^{q'(1+2s-sq)}\right)|f(x)|^{q's(q-2)},$$

and apply the Hölder–Rogers inequality with the pairwise conjugate exponents $p-1$ and $(p-1)/(p-2)$ in the first integral, and the exponents $q-1$ and $(q-1)/(q-2)$ in the second integral, respectively. This gives

$$\begin{aligned}\left(\int_{\mathbb{R}^n} |f(x)|\,dx\right)^{p+q} &\le C^{p+q}\\ \times \int_{\mathbb{R}^n} |x|^{n-\alpha}|f(x)|^{p(1+2r-rp)}\,dx &\int_{\mathbb{R}^n} |x|^{n+\alpha}|f(x)|^{q(1+2s-sq)}\,dx \qquad (5.21)\\ \times \left(\int_{\mathbb{R}^n} |f(x)|^{rp}\,dx\right)^{p-2} &\left(\int_{\mathbb{R}^n} |f(x)|^{sq}\,dx\right)^{q-2}.\end{aligned}$$

Since

$$1 - \frac{1-\theta}{p_0} - \frac{\theta}{p_1} = \frac{2}{p+q},$$

we can choose

$$C = C_0 A^{2/(p+q)} = C_0 \left(\frac{A_0}{\alpha}\right)^{2/(p+q)},$$

so that

$$C^{p+q} = \frac{D}{\alpha^2},$$

where D does not depend on α. In view of (5.16), we get

$$|x| \leq \left(\frac{g(x)}{M}\right)^{1/\gamma}.$$

Now, using this inequality to estimate the factors $|x|^{n-\alpha}$ and $|x|^{n+\alpha}$ in the integrals on the right-hand side of (5.21) yields (5.17), and the proof is complete. □

5.4.2 *An Extension of Theorem 5.8*

We show how Theorem 6.1, in the shape of Lemma 5.1, can be applied to get the following generalized versions of Theorem 5.8. We give a detailed proof of the continuous version only — the discrete version is proved similarly (see also [50]).

We define the Fourier transform $\hat{f}$ of an integrable function f on $\mathbb{R}^n$ by

$$\hat{f}(\omega) = \int_{\mathbb{R}^n} f(x) e^{-i\omega\cdot x}\, dx, \quad \omega \in \mathbb{R}^n.$$

Theorem 5.10 Suppose that $1 < r_0, r_1 \leq 2$, and that α is a positive integer satisfying

$$\alpha > \frac{n}{r_0}. \tag{5.22}$$

Define

$$\theta = \frac{\alpha - \frac{n}{r_0}}{\alpha + n\left(\frac{1}{r_1} - \frac{1}{r_0}\right)}.$$

Then there is a constant C such that

$$\|\hat{f}\|_{L_1(\mathbb{R}^n)} \leq C \left(\sum_{|\gamma|=\alpha} \|D^\gamma f\|_{L_{r_0}(\mathbb{R}^n)} \right)^{1-\theta} \|f\|^{\theta}_{L_{r_1}(\mathbb{R}^n)}.$$

Proof. Define, on $\mathbb{R}^n$, the measure

$$d\mu(\omega) = \frac{d\omega}{|\omega|^n}$$

and the weights

$$w(\omega) = |\omega|^n,$$

$$w_0(\omega) = |\omega|^{n/r_0'} \sum_{|\gamma|=\alpha} |\omega^\gamma|,$$

and

$$w_1(\omega) = |\omega|^{n/r_1'}.$$

Put $p = 1$, $p_0 = r_0'$ and $p_1 = r_1'$. Moreover, let

$$\begin{aligned} W(\omega) &= \frac{w(\omega)}{w_0^{1-\theta}(\omega) w_1^{\theta}(\omega)} \\ &= |\omega|^{n/q} \left(\sum_{|\gamma|=\alpha} |\omega^\gamma| \right)^{-(1-\theta)}, \end{aligned}$$

where q is defined by

$$\frac{1}{q} = 1 - \frac{1-\theta}{p_0} - \frac{\theta}{p_1}.$$

Then W is homogeneous of degree

$$\frac{n}{q} - (1-\theta)\alpha = 0,$$

and thus constant along rays from the origin. In particular, since W is continuous, as is easily seen, W is bounded on $\mathbb{R}^n$. This gives the condition

(5.19) of Lemma 5.1. Moreover, if we define V by

$$\begin{aligned}V(\omega) &= \frac{w_0(\omega)}{w_1(\omega)} \\ &= |\omega|^{n(\frac{1}{r_1}-\frac{1}{r_0})} \sum_{|\gamma|=\alpha} |\omega^\gamma|,\end{aligned}$$

then V is homogeneous of degree

$$\tau := \alpha + n\left(\frac{1}{r_1} - \frac{1}{r_0}\right),$$

and $\tau > 0$ by the assumption (5.22). Thus, there is a constant B_0 such that

$$\mu(\{2^m \le V < 2^{m+1}\}) \le \frac{B_0}{\tau}, \quad m \in \mathbb{Z},$$

which is (5.18). Therefore, by Lemma 5.1

$$\begin{aligned}&\|\hat{f}\|_{L_1(\mathbb{R}^n)} \\ &\le C\left(\int_{\mathbb{R}^n} |\hat{f}w_0|^{p_0}\, d\mu\right)^{(1-\theta)/p_0} \left(\int_{\mathbb{R}^n} |\hat{f}w_1|^{p_1}\, d\mu\right)^{\theta/p_1} \\ &= C\left(\int_{\mathbb{R}^n} \left(\sum_{|\gamma|=\alpha} |\hat{f}(\omega)\omega^\gamma|\right)^{r_0'} d\omega\right)^{(1-\theta)/r_0'} \left(\int_{\mathbb{R}^n} |\hat{f}(\omega)|^{r_1'}\, d\omega\right)^{\theta/r_1'} \\ &= C\left(\int_{\mathbb{R}^n} \left(\sum_{|\gamma|=\alpha} |\widehat{D^\gamma f}(\omega)|\right)^{r_0'} d\omega\right)^{(1-\theta)/r_0'} \left(\int_{\mathbb{R}^n} |\hat{f}(\omega)|^{r_1'}\, d\omega\right)^{(1-\theta)/r_1'} \\ &= C\left\|\sum |\gamma| = \alpha |\widehat{D^\gamma f}|\right\|_{L_{r_0'}(\mathbb{R}^n)}^{1-\theta} \|\hat{f}\|_{L_{r_1'}(\mathbb{R}^n)}^\theta \\ &\le C\left(\sum_{|\gamma|=\alpha} \|\widehat{D^\gamma f}\|_{L_{r_0'}(\mathbb{R}^n)}\right)^{1-\theta} \|\hat{f}\|_{L_{r_1'}(\mathbb{R}^n)}^\theta,\end{aligned}$$

where we have used the triangle inequality. Now, applying the Hausdorff–Young inequality (this is where we need the assumption $1 < r_i \le 2$) to each term in the sum and to the rightmost norm, we get

$$\|\hat{f}\|_{L_1(\mathbb{R}^n)} \le C\left(\sum_{|\gamma|=\alpha} \|D^\gamma f\|_{L_{r_0}(\mathbb{R}^n)}\right)^{1-\theta} \|f\|_{L_{r_1}(\mathbb{R}^n)}^\theta,$$

which is what we wanted to prove. □

In the discrete case, we need to treat the 0th Fourier coefficient with a little extra care. Theorem 5.8 follows if we put $r_0 = r_1 = r$ below.

Theorem 5.11 Suppose that $1 < r_0, r_1 \leq 2$, and that α is a positive integer such that

$$\alpha > \frac{n}{r_0}.$$

Define θ as in Theorem 5.10. Then there is a constant C such that

$$\|f\|_{A(\mathbb{T}^n)} \leq |\hat{f}(0)| + C \left(\sum_{|\gamma|=\alpha} \|D^\gamma f\|_{L_{r_0}(\mathbb{T}^n)} \right)^{1-\theta} \|f\|^{\theta}_{L_{r_1}(\mathbb{T}^n)}.$$

Sketch of proof. Assume first that $\hat{f}(0) = 0$. We can then proceed as in the proof of Theorem 5.10, but define the weights w_* and the functions W and V only on the integer lattice. The general case follows easily from this. □

Chapter 6

Some Carlson Type Inequalities for Weighted Lebesgue Spaces with General Measures

The setup for a Carlson type inequality can be generalized to a general measure space. This fact will be stated and proved in Section 6.1 (see Theorem 6.1). The most general result if this type when we have a product measure with n factors (for any $n \in \mathbb{Z}_+$) is stated in Section 6.3 (see Theorem 6.4). In order to avoid unnecessary technical details, we state and prove our results of this type in the two-factor case in Section 6.2.

6.1 The Basic Case

Suppose that (Ω, μ) is a measure space, on which positive, measurable weight functions w, w_0 and w_1 are defined. We would like to translate the conditions (5.6) and (5.7) to this abstract setting. Suppose, for simplicity, that $n = 1$ and $S = \{(1,0)\}$. Thus $\Omega = (0, \infty)$. With μ defined by

$$d\mu(x) = \frac{dx}{x},$$

the weights in Theorem 5.1 are

$$\begin{aligned} w(x) &= x^d, \\ w_i(x) &= x^{d_i} \quad i = 0, 1. \end{aligned}$$

The condition (5.6) could then be interpreted as

$$\frac{w(x)}{w_0^{1-\theta}(x) w_1^{\theta}(x)} = 1, \quad 0 < x < \infty.$$

Thus the quotient

$$\frac{w}{w_0^{1-\theta} w_1^{\theta}}$$

is in $L_\infty(\Omega, \mu)$. In the general case, we may allow a slightly weaker condition (see Theorem 6.1 below). The condition (5.7) means that the quotient

$$\frac{w_0(x)}{w_1(x)}$$

is not constant. The corresponding condition in Theorem 6.1 below says that the weights w_0 and w_1 may not be too close to each other on arbitrarily large sets (in a measure theoretic sense). We may also prove the inequality for a wider range of the parameter p, by using an interpolation technique.

The theorem we have in mind is the following result by L. Larsson [50] from 2003.

Theorem 6.1 Suppose that $p, p_0, p_1 \in (0, \infty]$ and $\theta \in (0, 1)$ are such that

$$\frac{1}{q} := \frac{1}{p} - \frac{1-\theta}{p_0} + \frac{\theta}{p_1} \geq 0. \tag{6.1}$$

Suppose, moreover, that there is an $s \in [q, \infty]$ for which

$$\frac{w}{w_0^{1-\theta} w_1^{\theta}} \in L_s(\Omega, \mu), \tag{6.2}$$

and that for some constant B we have

$$\mu\left(\left\{\omega \in \Omega; 2^m \leq \frac{w_0(\omega)}{w_1(\omega)} < 2^{m+1}\right\}\right) \leq B, \quad m \in \mathbb{Z}. \tag{6.3}$$

Then there is a constant C such that the Carlson type inequality

$$\|fw\|_{L_p(\Omega,\mu)} \leq C \|fw_0\|_{L_{p_0}(\Omega,\mu)}^{1-\theta} \|fw_1\|_{L_{p_1}(\Omega,\mu)}^{\theta} \tag{6.4}$$

holds for all measurable functions f.

Remark 6.1 As the proof of Theorem 6.1 will show, the condition (6.3) is not necessary if (6.2) holds with $s = q$, where q is defined by (6.1). However, there are examples showing that (6.3) is needed if we only assume that $w/w_0^{1-\theta} w_1^{\theta}$ is in L_s for some $s > q$ (see Remark 6.5 below).

Remark 6.2 The proof will also show that the constant C can be chosen to have the form

$$C = C_0 B^{1/q-1/s} \left\| \frac{w}{w_0^{1-\theta} w_1^{\theta}} \right\|_{L_s(\Omega,\mu)},$$

where C_0 does not depend on w, w_0, or w_1.

Remark 6.3 Theorem 6.1 applied with $p = 1$ and $s = \infty$, together with Remark 6.1, gives the crucial Lemma 5.1 of Chapter 5.

Let us give an example of an elementary but non-trivial inequality, where Theorem 6.1 needs to be applied with s strictly between q and ∞.

Proposition 6.1 Let $p_0, p_1 \geq 1$ and $\theta \in (0,1)$. Define q by

$$\frac{1}{q} = 1 - \frac{1-\theta}{p_0} - \frac{\theta}{p_1}.$$

Moreover, let a be a non-zero real number and $t > 1$. Then there is a constant C_0, which is independent of a and t, such that for all Lebesgue measurable functions f on $(0,\infty)$ it holds that

$$\begin{aligned}\int_0^1 \frac{|f(x)|}{x^{1/tq}}\,dx &+ \int_1^\infty \frac{|f(x)|}{x^{1/q}}\,dx \\ &\leq \frac{C_0}{|a|^{(t-1)/q(t+1)}} \max\{1,(t-1)^{-1/q}\} \\ &\times \left(\int_0^\infty |f(x)e^{\theta a x}|^{p_0}\,dx\right)^{(1-\theta)/p_0} \left(\int_0^\infty |f(x)e^{-(1-\theta)ax}|^{p_1}\,dx\right)^{\theta/p_1}.\end{aligned}$$

Proof. Let

$$w(x) = \begin{cases} x^{-1/tq}, & 0 < x < 1, \\ x^{-1/q}, & x \geq 1. \end{cases}$$

Put $w_0(x) = e^{\theta a x}$ and $w_1(x) = e^{-(1-\theta)ax}$. If $|E|$ denotes the Lebesgue measure of the set E, then

$$\begin{aligned}&\left|\left\{2^m \leq \frac{w_0}{w_1} < 2^{m+1}\right\}\right| \\ =&|\{x; 2^m \leq e^{ax} < 2^{m+1}\}| = \frac{1}{|a|}\log 2.\end{aligned}$$

This is (6.3) with

$$B = \frac{1}{|a|}\log 2.$$

Since $w_0^{1-\theta} w_1^{\theta} = 1$, we have

$$\frac{w}{w_0^{1-\theta} w_1^{\theta}} = w,$$

and $w \in L_s$ if and only if $q < s < tq$. Thus Theorem 6.1 applies. If

$$s = \frac{1+t}{2}q,$$

then

$$\|w\|_s = \left(\frac{2(t+1)}{t-1}\right)^{2/(t+1)q}.$$

This behaves like $4^{1/q}(t-1)^{-1/q}$ when t is close to 1 and tends to 1 as $t \to \infty$. By Remark 6.2, we can write

$$C = C_0 B^{1/q-1/s} \left\| \frac{w}{w_0^{1-\theta} w_1^{\theta}} \right\|_{L_s},$$

and with the choice of s made above we have

$$\frac{1}{q} - \frac{1}{s} = \frac{1}{q}\frac{t-1}{t+1},$$

and hence the desired result follows. □

Remark 6.4 We show by an example that the condition

$$\frac{1}{p} \geq \frac{1-\theta}{p_0} + \frac{\theta}{p_1} \tag{6.5}$$

of Theorem 6.1 cannot, in general, be relaxed. Consider $\Omega = [0,1]$ and $d\mu(x) = dx$. Fix $\theta \in (0,1)$ and $p_0, p_1 > 0$. Let $w = w_0 = w_1 = 1$. Since the measure is finite, the condition (6.3) holds trivially. Moreover,

$$W = \frac{w}{w_0^{1-\theta} w_1^{\theta}} = 1,$$

and hence $W \in L_s$ for any s. Let $\epsilon > 0$, and let f_ϵ be the characteristic function of the interval $[0, \epsilon]$. Then, for $w_* = 1$ and any $p_* > 0$,

$$\int_\Omega |f_\epsilon w_*|^{p_*} \, d\mu = \epsilon,$$

and hence

$$\frac{\|f_\epsilon w\|_{L_p}}{\|f_\epsilon w_0\|_{L_{p_0}}^{1-\theta} \|f_\epsilon w_1\|_{L_{p_1}}^{\theta}} = \epsilon^{\frac{1}{p} - \frac{1-\theta}{p_0} - \frac{\theta}{p_1}}.$$

Thus, if (6.5) is violated, this quotient can be made as large as we want by choosing ϵ small, i.e. there is no finite constant C such that (6.4) holds in this setting.

Let us show that the condition (6.3) cannot be removed in the case $s > q$ in order for (6.4) to hold in general. We consider $\Omega = (1, \infty)$ and μ defined by $d\mu(x) = dx$. Let

$$w(x) = x^{-\frac{1}{p}}$$

$$\text{and} \quad w_i(x) = x^{-\frac{1}{p_i}}, \quad i = 0, 1,$$

so that

$$\frac{w(x)}{w_0^{1-\theta}(x) w_1^{\theta}(x)} = x^{-1/q}.$$

This quotient is in L_s precisely when $s > q$. For $R > 1$, let f_R be the characteristic function of the interval $(1, R)$. Then, for any $w_*(x) = x^{-1/p_*}$ and any $p_* > 0$, we have

$$\int_\Omega |f_R w_*|^{p_*} \, d\mu = \int_1^R \frac{dx}{x} = \log R,$$

and thus

$$\frac{\|f_R w\|_{L_p}}{\|f_R w_0\|_{L_{p_0}}^{1-\theta} \|f_R w_1\|_{L_{p_1}}^{\theta}} = (\log R)^{1/q}.$$

When $R \to \infty$, this clearly tends to infinity and the claim follows.

Proof of Theorem 6.1. We may assume, without loss of generality, that $p_0 \leq p_1$. Moreover, we can assume that all parameters are ≥ 1 (this is needed when we apply the Riesz–Thorin interpolation theorem below). Consider the diagram in Figure 6.1. The strategy is to prove (6.4) first in the case when $s = \infty$ and $p \leq p_0, p_1$, after that on the "diagonal" $s = q$. We then apply an interpolation argument in order to get the inequality (6.4) in the convex hull of these sets (i.e. in the shaded region of the diagram in Figure 6.1). This will complete the proof.

Thus, suppose first that

$$\frac{w}{w_0^{1-\theta} w_1^{\theta}} \in L_\infty(\Omega, \mu)$$

and $p \leq p_0$. For $i = 0, 1$, define r_i by

$$\frac{1}{r_i} = \frac{1}{p} - \frac{1}{p_i}.$$

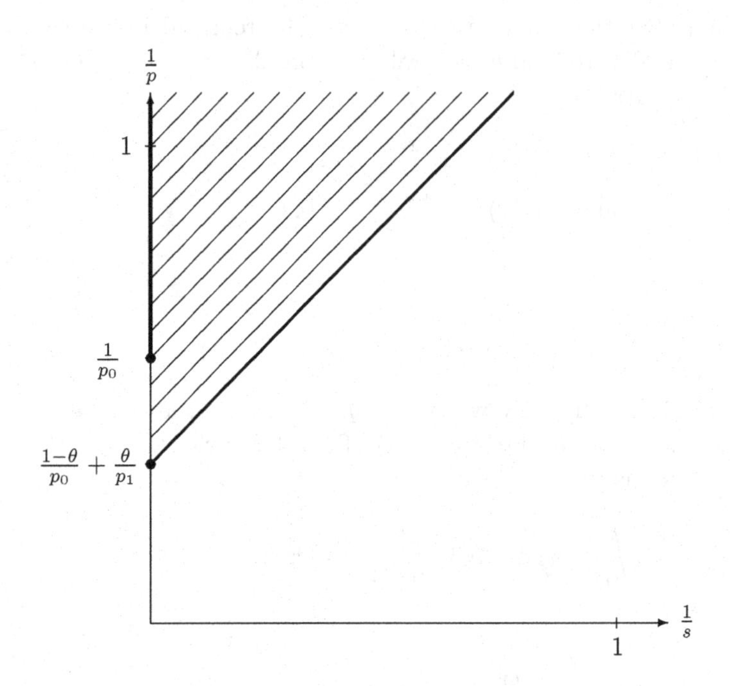

Fig. 6.1 The diagram shows the region of admissible parameters for a Carlson type inequality on a general measure space.

It follows by the Hölder–Rogers inequality that if E is any measurable subset of Ω, then

$$\begin{aligned}
\int_\Omega |fw|^p \, d\mu &= \int_{\Omega \setminus E} \left(\frac{w}{w_0}\right)^p |fw_0|^p \, d\mu + \int_E \left(\frac{w}{w_1}\right)^p |fw_1|^p \, d\mu \\
&\leq \left(\int_{\Omega \setminus E} \left(\frac{w}{w_0}\right)^{r_0} d\mu\right)^{p/r_0} \left(\int_\Omega |fw_0|^{p_0} \, d\mu\right)^{p/p_0} \\
&\quad + \left(\int_E \left(\frac{w}{w_1}\right)^{r_1} d\mu\right)^{p/r_1} \left(\int_\Omega |fw_1|^{p_1} \, d\mu\right)^{p/p_1} \\
&= M_0^p \, \|fw_0\|_{L_{p_0}(\Omega,\mu)}^p + M_1^p \, \|fw_1\|_{L_{p_1}(\Omega,\mu)}^p .
\end{aligned}$$

Let δ be any positive number. We want to choose the set E so as to obtain the estimates

$$M_0 \le N_0 \delta^\theta \quad \text{and} \quad M_1 \le N_1 \delta^{-(1-\theta)}, \tag{6.6}$$

where N_0 and N_1 are some constants. For if we let

$$\delta = \left(\frac{1-\theta}{\theta}\right)^{1/p} \frac{N_1 \|fw_1\|_{L_{p_1}(\Omega,\mu)}}{N_0 \|fw_0\|_{L_{p_0}(\Omega,\mu)}},$$

this yields (6.4) with

$$C = \frac{N_0^{1-\theta} N_1^\theta}{((1-\theta)^{1-\theta}\theta^\theta)^{1/p}}.$$

For $m \in \mathbb{Z}$, we define Ω_m to be the subset of Ω on which

$$2^m \le \frac{w_0}{w_1} < 2^{m+1}.$$

Moreover, we let E be the set where

$$\frac{w_0}{w_1} \ge 2^{m_0},$$

where the integer m_0 will be specified later on. It follows that

$$\begin{aligned}
M_0^{r_0} &= \int_{\Omega \setminus E} \left(\frac{w}{w_0^{1-\theta} w_1^\theta}\right)^{r_0} \left(\frac{w_0}{w_1}\right)^{\theta r_0} d\mu \\
&= \sum_{m=-\infty}^{m_0-1} \int_{\Omega_m} \left(\frac{w}{w_0^{1-\theta} w_1^\theta}\right)^{r_0} \left(\frac{w_0}{w_1}\right)^{\theta r_0} d\mu \\
&\le \left\|\frac{w}{w_0^{1-\theta} w_1^\theta}\right\|_{L_\infty(\Omega,\mu)}^{r_0} B \sum_{m=-\infty}^{m_0-1} 2^{(m+1)\theta r_0} \\
&= \left\|\frac{w}{w_0^{1-\theta} w_1^\theta}\right\|_{L_\infty(\Omega,\mu)}^{r_0} B \frac{2^{\theta m_0 r_0}}{1-2^{-\theta r_0}},
\end{aligned} \tag{6.7}$$

and similarly that

$$M_1^{r_1} \le \left\|\frac{w}{w_0^{1-\theta} w_1^\theta}\right\|_{L_\infty(\Omega,\mu)}^{r_1} B \frac{2^{-(1-\theta) m_0 r_1}}{1-2^{-(1-\theta) r_1}}. \tag{6.8}$$

Let

$$W = \left\|\frac{w}{w_0^{1-\theta} w_1^\theta}\right\|_{L_\infty(\Omega,\mu)},$$

and suppose that the constants N_0 and N_1 are chosen so that

$$\left(\frac{N_0}{WB^{1/r_0}}\right)^{\frac{1}{\theta}}(1-2^{-\theta r_0})^{1/\theta r_0}=2\left(\frac{N_1}{WB^{1/r_1}}\right)^{-\frac{1}{1-\theta}}(1-2^{-(1-\theta)r_1})^{-1/(1-\theta)r_1}. \tag{6.9}$$

We can then choose m_0 so that

$$\left(\frac{N_1}{WB^{1/r_1}}\right)^{-\frac{1}{1-\theta}}(1-2^{-(1-\theta)r_1})^{-1/(1-\theta)r_1}\leq\frac{2^{m_0}}{\delta}\leq\left(\frac{N_0}{WB^{1/r_0}}\right)^{\frac{1}{\theta}}(1-2^{-\theta r_0})^{1/\theta r_0}$$

which together with (6.7) and (6.8) gives (6.6). It is readily shown that (6.9) is equivalent to

$$N_0^{1-\theta}N_1^{\theta}=\frac{2^{\theta(1-\theta)}WB^{\frac{1}{q}}}{(1-2^{-\theta r_0})^{(1-\theta)/r_0}(1-2^{-(1-\theta)r_1})^{\theta/r_1}}.$$

Thus, in view of the discussions above, we have proved that (6.4) holds with

$$C=\left\|\frac{w}{w_0^{1-\theta}w_1^{\theta}}\right\|_{L_\infty(\Omega,\mu)}\frac{2^{\theta(1-\theta)}B^{\frac{1}{q}}}{((1-\theta)^{1-\theta}\theta^{\theta})^{\frac{1}{p}}(1-2^{-\theta r_0})^{(1-\theta)/r_0}(1-2^{-(1-\theta)r_1})^{\theta/r_1}}.$$

For later purposes, we write

$$C_0=\frac{2^{\theta(1-\theta)}B^{\frac{1}{q}}}{((1-\theta)^{1-\theta}\theta^{\theta})^{\frac{1}{p}}(1-2^{-\theta r_0})^{(1-\theta)/r_0}(1-2^{-(1-\theta)r_1})^{\theta/r_1}}, \tag{6.10}$$

so that (6.4) holds with

$$C=C_0\left\|\frac{w}{w_0^{1-\theta}w_1^{\theta}}\right\|_{L_\infty(\Omega,\mu)}.$$

Suppose next that $s=q$. This can be written as

$$\frac{p}{s}+\frac{p(1-\theta)}{p_0}+\frac{p\theta}{p_1}=1,$$

and, hence, the Hölder–Rogers inequality can be applied with three factors, using the exponents

$$\frac{s}{p},\frac{p_0}{p(1-\theta)},\frac{p_1}{p\theta},$$

which yields

$$\begin{aligned}\int_\Omega |fw|^p \, d\mu &= \int_\Omega \left(\frac{w}{w_0^{1-\theta} w_1^\theta}\right)^p |fw_0|^{p(1-\theta)} |fw_1|^{p\theta} \, d\mu \\ &\le \left\| \frac{w}{w_0^{1-\theta} w_1^\theta} \right\|_{L_q(\Omega,\mu)}^p \|fw_0\|_{L_{p_0}(\Omega,\mu)}^{p(1-\theta)} \|fw_1\|_{L_{p_1}(\Omega,\mu)}^{p\theta} .\end{aligned}$$

This is precisely (6.4) with

$$C = \left\| \frac{w}{w_0^{1-\theta} w_1^\theta} \right\|_{L_q(\Omega,\mu)} .$$

It remains to prove the inequality for $(\frac{1}{s}, \frac{1}{p})$ in the shaded region in Figure 6.1. Fix the function f, the weights w_0 and w_1, and the parameters p_0 and p_1. Define the linear operator

$$T : L_s(\Omega, \mu) \to L_p(\Omega, \mu)$$

by

$$T\omega = (fw_0^{1-\theta} w_1^\theta)\omega.$$

For any $\omega \in L_s(\Omega, \mu)$, we can write

$$|\omega| = \frac{w}{w_0^{1-\theta} w_1^\theta}$$

for the correct choice of w. The previously shown cases of (6.4) now state that

$$\|T\omega\|_{L_p(\Omega,\mu)} \le C_0 \|fw_0\|_{L_{p_0}(\Omega,\mu)}^{1-\theta} \|fw_1\|_{L_{p_1}(\Omega,\mu)}^\theta \|\omega\|_{L_\infty(\Omega,\mu)}$$

and

$$\|T\omega\|_{L_p(\Omega,\mu)} \le \|fw_0\|_{L_{p_0}(\Omega,\mu)}^{1-\theta} \|fw_1\|_{L_{p_1}(\Omega,\mu)}^\theta \|\omega\|_{L_q(\Omega,\mu)} ,$$

respectively, where C_0 is as defined in (6.10) with any admissible p_0 and p_1. In other words, T is a bounded linear operator $L_\infty \to L_p$ with norm at most

$$C_0 \|fw_0\|_{L_{p_0}(\Omega,\mu)}^{1-\theta} \|fw_1\|_{L_{p_1}(\Omega,\mu)}^\theta$$

and $L_q \to L_p$ with norm not exceeding

$$\|fw_0\|_{L_{p_0}(\Omega,\mu)}^{1-\theta} \|fw_1\|_{L_{p_1}(\Omega,\mu)}^\theta .$$

Thus, if $0 < \sigma < 1$, then the Riesz–Thorin interpolation theorem (see e.g. [12] or [18]) implies the boundedness of $T : L_{s_\sigma} \to L_p$, where s_σ is defined by the relation

$$\frac{1}{s_\sigma} = \frac{1-\sigma}{\infty} + \frac{\sigma}{q} = \frac{\sigma}{q},$$

with norm at most

$$C_0^{1-\sigma} \|fw_0\|_{L_{p_0}(\Omega,\mu)}^{1-\theta} \|fw_1\|_{L_{p_1}(\Omega,\mu)}^{\theta}.$$

By translating this fact back to the original situation, this shows that (6.4) holds also in the remaining cases, completing the proof. □

6.2 The Product Measure Case – Two Factors

Once we have Theorem 6.1, a corresponding Carlson type inequality for product measure spaces can be proved. Before we present our most general result (see Theorem 6.4 in Section 6.3), we will in this section clarify our ideas by stating and proving two versions of the two-factor case.

As before, we define q by

$$\frac{1}{q} = \frac{1}{p} - \frac{1-\theta}{p_0} - \frac{\theta}{p_1}, \tag{6.11}$$

and we assume that this quantity is ≥ 0.

Theorem 6.2 Suppose, in addition to the assumptions of Theorem 6.1, that (Ξ, ν) is a measure space with weights v, v_0, and v_1, and assume that

$$\frac{v}{v_0^{1-\theta} v_1^{\theta}} \in L_q(\nu). \tag{6.12}$$

Let $X = \Omega \times \Xi$ and $d\kappa = d\mu \times d\nu$, and put

$$\begin{aligned} u(\omega, \xi) &= w(\omega)v(\xi) \\ \text{and} \quad u_i(\omega, \xi) &= w_i(\omega)v_i(\xi), i = 0, 1, \quad \omega \in \Omega, \xi \in \Xi. \end{aligned}$$

Then

$$\|fu\|_{L_p(X,\kappa)} \leq C' \|fu_0\|_{L_{p_0}(X,\kappa)}^{1-\theta} \|fu_1\|_{L_{p_1}(X,\kappa)}^{\theta} \tag{6.13}$$

holds, where we may choose a constant of the form

$$C' = C_0 \left\| \frac{w}{w_0^{1-\theta} w_1^{\theta}} \right\|_{L_q(\Omega,\mu)} \left\| \frac{v}{v_0^{1-\theta} v_1^{\theta}} \right\|_{L_s(\Xi,\nu)},$$

where C_0 is independent of w, v, w_0, v_0, w_1 and v_1.

Proof. We assume that $p, p_0, p_1 < \infty$. Then (6.11) can be written as

$$\frac{p}{q} + \frac{(1-\theta)p}{p_0} + \frac{\theta p}{p_1} = 1,$$

and, therefore, the Hölder–Rogers inequality with three factors can be applied, using the exponents

$$\frac{q}{p}, \frac{p_0}{(1-\theta)p}, \frac{p_1}{\theta p}.$$

Assuming that (6.4) of Theorem 6.1 holds with

$$C = C_0 \left\| \frac{w}{w_0^{1-\theta} w_1^{\theta}} \right\|_{L_s(\Omega,\mu)},$$

we thus get, also using Fubini's theorem

$$\begin{aligned}
\|fu\|_{L_p(\kappa)}^p &= \int_\Xi v^p \int_\Omega |fw|^p \, d\mu \, d\nu \\
&\le C^p \int_\Xi \left(\frac{v}{v_0^{1-\theta} v_1^{\theta}} \right)^p \\
&\quad \left(\int_\Omega |fv_0 w_0|^{p_0} \, d\mu \right)^{(1-\theta)p/p_0} \left(\int_\Omega |fv_1 w_1|^{p_1} \, d\mu \right)^{\theta p/p_1} d\nu \\
&\le C^p \left(\int_\Xi \left(\frac{v}{v_0^{1-\theta} v_1^{\theta}} \right)^q d\nu \right)^{p/q} \\
&\quad \left(\int_\Xi \int_\Omega |fv_0 w_0|^{p_0} \, d\mu \, d\nu \right)^{(1-\theta)p/p_0} \\
&\quad \left(\int_\Xi \int_\Omega |fv_1 w_1|^{p_1} \, d\mu \, d\nu \right)^{\theta p/p_1} \\
&= C^p \left\| \frac{v}{v_0^{1-\theta} v_1^{\theta}} \right\|_{L_q(\Xi,\nu)}^p \|fu_0\|_{L_{p_0}(X,\kappa)}^{(1-\theta)p} \|fu_1\|_{L_{p_1}(X,\kappa)}^{\theta p}.
\end{aligned}$$

Taking pth roots, we get the desired inequality with

$$C' = C_0 \left\| \frac{v}{v_0^{1-\theta} v_1^{\theta}} \right\|_{L_q(\Xi,\nu)} \left\| \frac{w}{w_0^{1-\theta} w_1^{\theta}} \right\|_{L_s(\Omega,\mu)}.$$

If $p = \infty$, then we must also have $p_0 = p_1 = q = \infty$, and

$$|fu| \leq \left\| \frac{v}{v_0^{1-\theta} v_1^{\theta}} \right\|_{L_\infty(\Xi,\nu)} \left\| \frac{w}{w_0^{1-\theta} w_1^{\theta}} \right\|_{L_\infty(\Omega,\mu)} |fu_0v_0|^{1-\theta} |fu_1v_1|^{\theta}.$$

By taking suprema, the desired result follows (with $C_0 = 1$). If $p < \infty$ but p_0 or p_1 is finite, the inequality follows similarly. □

Theorem 6.2 is not symmetric, in the sense that we have different conditions on the respective measure spaces. If we impose a condition corresponding to (6.3) also on the second measure space factor, then we can loosen the condition (6.12) slightly, and we get a symmetric version of this two-dimensional result.

Theorem 6.3 Suppose that the hypotheses of Theorem 6.1 hold with $s = s_1$. Suppose, moreover, that

$$\nu(\{2^m \leq v_0/v_1 < 2^{m+1}\}) \leq B, \quad m \in \mathbb{Z},$$

and that

$$\frac{v}{v_0^{1-\theta} v_1^{\theta}} \in L_{s_2}(\Xi, \nu),$$

where

$$0 \leq \frac{1}{s_1} \leq \frac{1}{q}, \quad 0 \leq \frac{1}{s_2} \leq \frac{1}{q}, \quad \frac{1}{s_1} + \frac{1}{s_2} \geq \frac{1}{q}. \tag{6.14}$$

Then the inequality (6.13) holds with

$$C' = C_0 \left\| \frac{w}{w_0^{1-\theta} w_1^{\theta}} \right\|_{L_{s_1}(\Omega,\mu)} \left\| \frac{v}{v_0^{1-\theta} v_1^{\theta}} \right\|_{L_{s_2}(\Xi,\nu)}.$$

Proof. Theorem 6.2 applied first as it is and then with the two measure space factors interchanged, gives the result when the point

$$\left(\frac{1}{s_1}, \frac{1}{s_2} \right)$$

is situated on two of the edges of the triangle shown in Figure 6.2. To prove the inequality for the remaining points under consideration (see the shaded area in the diagram of Figure 6.2), we use bilinear interpolation.

We now fix everything except the weights w and v, and define a bilinear operator

$$T : L_{s_2}(\Xi, \nu) \times L_{s_1}(\Omega, \mu) \to L_p(X, \kappa)$$

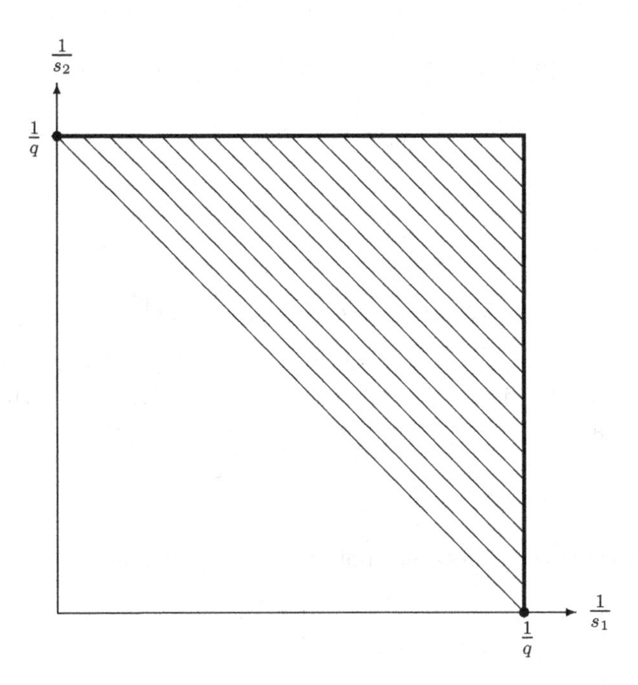

Fig. 6.2 For points (s_1^{-1}, s_2^{-1}) on the right and top edges of the shaded triangle in the diagram above, the conclusion of Theorem 6.3 is implied directly by Theorem 6.2 by letting the two factors switch roles. In the convex hull, then, the inequality follows by applying bilinear interpolation.

by

$$T(W, V) = [(fw_0v_0)^{1-\theta}(fw_1v_1)^{\theta}]WV.$$

Since

$$T\left(\frac{w}{w_0^{1-\theta}w_1^{\theta}}, \frac{v}{v_0^{1-\theta}v_1^{\theta}}\right) = fwv,$$

the inequality (6.13) with

$$C' = C_0 \left\| \frac{w}{w_0^{1-\theta}w_1^{\theta}} \right\|_{L_{s_1}(\Omega,\mu)} \left\| \frac{v}{v_0^{1-\theta}v_1^{\theta}} \right\|_{L_{s_2}(\Xi,\nu)}$$

can (with the appropriate interpretation of the underlying measure spaces) be written as

$$\|T(W,V)\|_p \leq C_0 \|fw_0v_0\|_{p_0}^{1-\theta} \|fw_1v_1\|_{p_1}^{\theta} \left\| \frac{v}{v_0^{1-\theta}v_1^{\theta}} \right\|_{s_2} \left\| \frac{w}{w_0^{1-\theta}w_1^{\theta}} \right\|_{s_1},$$

or, if we denote by $\|\cdot\|_{(s_1,s_2),p}$ the norm of the operator T, it holds that

$$\|T\|_{(\sigma_1,q),p} \leq C_0^{(1)} \|fw_0v_0\|_{p_0}^{1-\theta} \|fw_1v_1\|_{p_1}^{\theta}$$

and

$$\|T\|_{(q,\sigma_2),p} \leq C_0^{(2)} \|fw_0v_0\|_{p_0}^{1-\theta} \|fw_1v_1\|_{p_1}^{\theta}$$

whenever $\sigma_1, \sigma_2 \in [q,\infty]$. Suppose now that (s_1,s_2) is any point on the triangle (6.14) off the right and top edges (cf. Figure 6.2). Then there are $\sigma_1, \sigma_2 \in [q,\infty]$ and $\eta \in (0,1)$ such that

$$\frac{1}{s_1} = \frac{1-\eta}{\sigma_1} + \frac{\eta}{q} \quad \text{and} \quad \frac{1}{s_2} = \frac{1-\eta}{q} + \frac{\eta}{\sigma_2}.$$

Thus, by applying multi-linear interpolation (see e.g. Theorem 4.4.1 of [12]), it holds that

$$\begin{aligned}\|T\|_{(s_1,s_2),p} &\leq \left(C_0^{(1)} \|fw_0v_1\|_{p_0}^{1-\theta} \|fw_1v_1\|_{p_1}^{\theta}\right)^{1-\eta} \\ &\quad \left(C_0^{(2)} \|fw_0v_1\|_{p_0}^{1-\theta} \|fw_1v_1\|_{p_1}^{\theta}\right)^{\eta} \\ &= (C_0^{(1)})^{1-\eta}(C_0^{(2)})^{\eta} \|fw_0v_0\|_{p_0}^{1-\theta} \|fw_1v_1\|_{p_1}^{\theta},\end{aligned}$$

which, in turn, can be written as (6.13) with

$$C' = (C_0^{(1)})^{1-\eta}(C_0^{(2)})^{\eta} \left\| \frac{w}{w_0^{1-\theta}w_1^{\theta}} \right\|_{s_1} \left\| \frac{v}{v_0^{1-\theta}v_1^{\theta}} \right\|_{s_2}.$$

The proof is complete. □

Remark 6.6 The triangle (6.14) is, at least in the case $p_0 = p_1$, the largest possible region in which we have the inequality (6.13). It suffices to show failure on the diagonal $s_1 = s_2 = s$, $s > 2q$. To see this, consider the point P in Figure 6.3. By interpolation techniques, we know that the region in which we have the inequality (6.13) is convex. Thus, if we show that such an inequality does not hold on the dashed diagonal, then it cannot hold at the point P either, since we can draw a straight line segment joining P to the triangle defined by (6.14) crossing the diagonal. It only remains to prove that Theorem 6.3 fails in general on the dashed diagonal in the diagram in Figure 6.3.

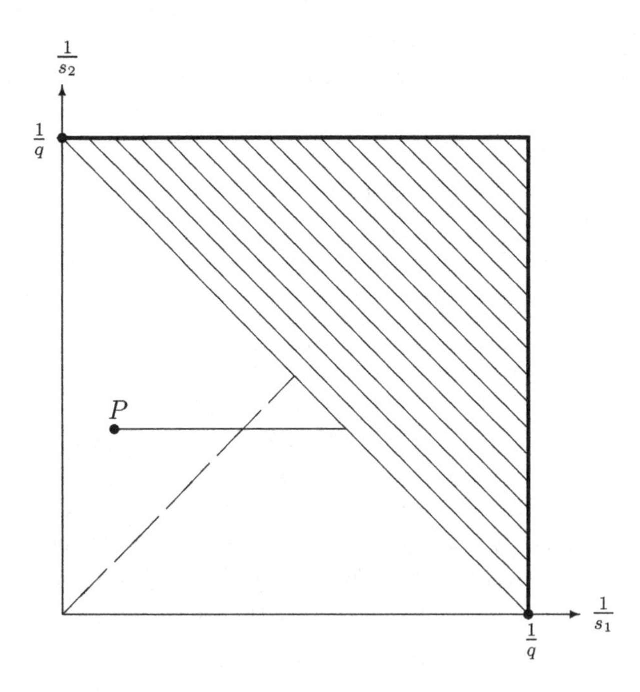

Fig. 6.3 To show failure of the Carlson type inequality on the product measure space, it suffices to consider the diagonal, since the region in which inequality holds is necessarily convex, as is seen by interpolation.

To do this, we let $\Omega = \Xi = (2, \infty)$ and consider the measures μ and ν defined by

$$d\mu(x) = \frac{dx}{x} \quad \text{and} \quad d\nu(y) = \frac{dy}{y}.$$

Define

$$w_0(x) = x^{1/p_0}, \quad w_1(x) = x^{1+1/p_0},$$

$$v_0(y) = y^{1+1/p_0}, \quad v_1(y) = y^{1/p_0}.$$

Also, let

$$w = g w_0^{1-\theta} w_1^{\theta}$$

and

$$v = g v_0^{1-\theta} v_1^{\theta},$$

where

$$g(t) = (\log t)^{-1/2q}.$$

Then

$$V = \frac{v}{v_0^{1-\theta} v_1^{\theta}}$$

and

$$W = \frac{w}{w_0^{1-\theta} w_1^{\theta}}$$

are both in L_s if and only if $s > 2q$. The condition (6.3) is satisfied on both spaces, and hence Theorem 6.1 guarantees that the Carlson type inequality holds on both factors. Consider, however, the functions f_R, defined on $\Omega \times \Xi$ by

$$f_R(x,y) = (g(x)g(y))^{q/p_0} \sqrt{x^2+y^2}^{-1-2/p_0} K_R(x,y),$$

where K_R is the characteristic function of the set

$$P_R = \left\{ (x,y) \in (2,\infty)^2; r_0 \le \sqrt{x^2+y^2} \le R, \frac{\pi}{8} \le \arctan \frac{y}{x} \le \frac{3\pi}{8} \right\},$$

and

$$r_0^2 = \frac{8\sqrt{2}}{\sqrt{2}-1}.$$

It can be shown that there are constants d and D such that for all $(x,y) \in P_R$

$$d(\log \sqrt{x^2+y^2})^2 \le (\log x)(\log y) \le D(\log \sqrt{x^2+y^2})^2.$$

Thus if κ denotes the product measure, we have

$$\int |f_R v w|^p \, d\kappa \ge c \int_{r_0}^{R} \frac{dr}{r \log r}$$

and

$$\left(\int |f_R v_0 w_0|^{p_0}\, d\kappa\right)^{(1-\theta)/p_0}$$
$$\times\left(\int |f_R v_1 w_1|^{p_0}\, d\kappa\right)^{\theta/p_0} \le C\left(\int_{r_0}^{R} \frac{dr}{r\log r}\right)^{1/p_0}.$$

It follows that

$$\frac{\|f_R vw\|_p}{\|f_R v_0 w_0\|_{p_0}^{1-\theta}\,\|f_R v_1 w_1\|_{p_0}^{\theta}}$$
$$\ge H\left(\int_{r_0}^{R} \frac{dr}{r\log r}\right)^{1/q} \to \infty, \quad R\to\infty,$$

which proves the statement.

6.3 The General Product Measure Case

We conclude this chapter with a brief discussion of Carlson type inequalities on product measure spaces with any finite number of factors.

For $j = 1,\ldots,n$, let $(\Omega_j, \mu^{(j)})$ be a σ-finite measure space, on which weights $w^{(j)}$, $w_0^{(j)}$ and $w_1^{(j)}$ are defined. We let

$$\Omega = \Omega_1 \times \ldots \times \Omega_n$$

and

$$d\mu = d\mu_1 \times \ldots \times d\mu_n,$$

and define w on Ω by

$$w(\omega_1,\ldots,\omega_n) = w^{(1)}(\omega_1)\cdots w^{(n)}(\omega_n);$$

w_0 and w_1 are defined analogously. Furthermore, we put

$$W^{(j)} = \frac{w^{(j)}}{(w_0^{(j)})^{1-\theta}(w_1^{(j)})^{\theta}}, \quad j = 1,\ldots,n.$$

We then have the following multi-dimensional extension of Theorem 6.1, whose proof can be found in [50].

Theorem 6.4 Let k be an integer such that $0 \leq k \leq n$. Suppose that

$$W^{(j)} \in L_{s_j}(\Omega_j, \mu^{(j)}), \quad j = 1, \ldots, k,$$

where

$$0 \leq \frac{1}{s_j} \leq \frac{1}{q}, \quad \frac{1}{s_1} + \ldots + \frac{1}{s_k} \geq \frac{k-1}{q}, \tag{6.15}$$

and

$$W^{(j)} \in L_q(\Omega_j, \mu^{(j)}), \quad j = k+1, \ldots, n.$$

Suppose, moreover, that for $j = 1, \ldots, k$, there are constants B^j such that

$$\mu^{(j)}(\{2^m \leq w_0^{(j)}/w_1^{(j)} < 2^{m+1}\}) \leq B_j, \quad m \in \mathbb{Z}.$$

Then the inequality

$$\|f\|_{L_p(\Omega, w^p \mu)} \leq C \|f\|^{1-\theta}_{L_{p_0}(\Omega, w_0^{p_0} \mu)} \|f\|^{\theta}_{L_{p_1}(\Omega, w_1^{p_1} \mu)} \tag{6.16}$$

holds for some constant C.

Remark 6.7 The example in Remark 6.6 above can be generalized to n factors, but only to a smaller set than the complement of (6.15) if $k > 2$. It remains an open question whether the region defined by (6.15) is the largest possible set on which we can prove the inequality (6.16) in general.

Chapter 7

Carlson Type Inequalities and Real Interpolation Theory

In the previous chapter, we saw that interpolation theory can be used to prove some new Carlson type inequalities. The aim of the present chapter is to initiate a discussion of a closer connection between Carlson type inequalities and real interpolation theory. More information on this theme can be found in Chapter 8.

In order to make our text reasonably self-contained, we present in Sections 7.1 and 7.2 notation and basic facts concerning interpolation of normed spaces and the real interpolation method, respectively.

In Section 7.3, we state and prove a recent result by L. Larsson [51], concerning embeddings of real interpolation spaces into weighted Lebesgue spaces.

Throughout this chapter, we assume that A_0 and A_1 are normed spaces, which are *compatible*, in the sense that they are continuously embedded in a Hausdorff topological vector space $\mathcal{V}$. We denote by $\bar{A}$ the couple (A_0, A_1) of normed spaces. For notational simplicity, the norms on A_0 and A_1 will sometimes be denoted by $\|\cdot\|_0$ and $\|\cdot\|_1$, respectively. Since the results of purely interpolation theoretical nature can be found in numerous texts on the matter (see e.g. [12] or [18]), the ones we will need are presented here without proofs. If the two spaces A_0 and A_1 are Banach spaces, we will call $\bar{A}$ a *Banach couple*.

7.1 Interpolation of Normed Spaces

If A_0 and A_1 are compatible, it makes sense to define the *intersection* and *sum* of $\bar{A}$ by

$$\Delta(\bar{A}) = A_0 \cap A_1$$

and

$$\Sigma(\bar{A}) = \{f_0 + f_1; f_i \in A_i, i = 0, 1\},$$

respectively. Here, of course, $+$ denotes addition in $\mathcal{V}$.

$\Delta(\bar{A})$ and $\Sigma(\bar{A})$ are normed spaces under the norms defined by

$$\|f\|_{\Delta(\bar{A})} = \max\{\|f\|_0, \|f\|_1\}$$

and

$$\|f\|_{\Sigma(\bar{A})} = \inf\{\|f_0\|_0 + \|f_1\|_1; f = f_0 + f_1, f_i \in A_i, i = 0, 1\},$$

respectively. Moreover, they are Banach spaces if A_0 and A_1 are.

Suppose that X is a normed subspace of $\mathcal{V}$. X is then called an *intermediate space between A_0 and A_1* if

$$\Delta(\bar{A}) \subseteq X \subseteq \Sigma(\bar{A}).$$

Recall that this means that the inclusions are continuous.

We write $T : \bar{A} \to \bar{A}$ to denote a linear mapping T on $\Sigma(\bar{A})$ such that

$$T|_{A_i} : A_i \to A_i, \quad i = 0, 1;$$

thus we say that T is continuous on $\bar{A}$ if T maps the subspaces A_i continuously into themselves. An intermediate space X between A_0 and A_1 will be called an *interpolation space between A_0 and A_1* if it has the additional property that

$$T|_X : X \to X$$

whenever $T : \bar{A} \to \bar{A}$.

7.2 The Real Interpolation Method

There are many known methods to produce interpolation spaces. In this section, we will present two of the most well-known methods (the K-method and the J-method), which turn out to be equivalent. Any of the two methods is then called the real interpolation method. There are also other equivalent characterizations of this method, which is very useful for several applications.

7.2.1 *The K-method*

If $t > 0$, we define the *Peetre K-functional* on $\Sigma(\bar{A})$ by

$$K(t, f; \bar{A}) = \inf\{\|f_0\|_0 + t\,\|f_1\|_1\,; f = f_0 + f_1, f_i \in A_i, i = 0, 1\}.$$

For any fixed $t > 0$, $K(t, \cdot; \bar{A})$ is a norm on $\Sigma(\bar{A})$, which is equivalent to the existing norm on $\Sigma(\bar{A})$ (note that the latter is $K(1, \cdot; \bar{A})$).

Definition 7.1 Suppose that $0 < \theta < 1$ and $1 \le p \le \infty$. We define $\bar{A}_{\theta,p}$ to be the space of $f \in \Sigma(\bar{A})$ for which

$$\|f\|_{\theta,p} = \begin{cases} \left(\int_0^\infty (t^{-\theta} K(t, f; \bar{A}))^p \frac{dt}{t}\right)^{1/p}, & 1 \le p < \infty, \\ \sup_{t>0} t^{-\theta} K(t, f; \bar{A}), & p = \infty \end{cases}$$

is finite.

Proposition 7.1 For $0 < \theta < 1$ and $1 \le p \le \infty$, the space $\bar{A}_{\theta,p}$ is an interpolation space between A_0 and A_1.

7.2.2 *The J-method*

If $t > 0$, we define the *Peetre J-functional* on $\Delta(\bar{A})$ by

$$J(t, f; \bar{A}) = \max\{\|f\|_0, t\,\|f\|_1\}.$$

The existing norm on $\Delta(\bar{A})$ is $J(1, \cdot; \bar{A})$. Any other $t > 0$ gives an equivalent norm on $\Delta(\bar{A})$.

Definition 7.2 Suppose that $0 < \theta < 1$ and $1 \le p \le \infty$. We define $\bar{A}^J_{\theta,p}$ as the space of those f in $\Sigma(\bar{A})$ which can be written as

$$f = \int_0^\infty u(t) \frac{dt}{t} \tag{7.1}$$

for some strongly measurable function $u : \mathbb{R}_+ \to \Delta(\bar{A})$. We define a norm on $\bar{A}^J_{\theta,p}$ by taking the infimum of the expressions

$$\begin{cases} \left(\int_0^\infty (t^{-\theta} J(t, u(t); \bar{A}))^p \frac{dt}{t}\right)^{1/p}, & 1 \le p < \infty, \\ \sup_{t>0} t^{-\theta} J(t, u(t); \bar{A}), & p = \infty, \end{cases}$$

where u ranges over all the functions for which (7.1) holds.

Proposition 7.2 For $0 < \theta < 1$ and $1 \le p \le \infty$, the space $\bar{A}^J_{\theta,p}$ is an interpolation space between A_0 and A_1.

7.2.3 *The Equivalence Theorem*

The K- and J-methods are equivalent, in the sense that they produce interpolation spaces which, as topological vector spaces, are equal.

Proposition 7.3 (The Equivalence Theorem) Given the couple $\bar{A} = (A_0, A_1)$ of normed spaces, $\theta \in (0,1)$ and $p \in [1, \infty]$, we have

$$\bar{A}^J_{\theta,p} = \bar{A}_{\theta,p},$$

with equivalent norms.

Any of the two equivalent K- and J-methods is referred to as *the real interpolation method*, and the resulting space is called a *real interpolation space*.

7.2.4 *The Classes $\mathcal{C}_J$ and $\mathcal{C}_K$*

Given the couple $\bar{A}$ and $\theta \in (0,1)$, an intermediate space X is said to be of class $\mathcal{C}_J = \mathcal{C}_J(\theta; \bar{A})$ if there is a constant C such that for all $f \in \Delta(\bar{A})$ it holds that

$$\|f\|_X \le C t^{-\theta} J(t, f; \bar{A}). \tag{7.2}$$

The following result relates the real interpolation method to Carlson type inequalities.

Proposition 7.4 If X is a Banach space, then the following statements are equivalent:

(i) X is of class $\mathcal{C}_J(\bar{A})$.
(ii) $\bar{A}_{\theta,1} \subseteq X$.
(iii) The Carlson type inequality

$$\|f\|_X \le C \, \|f\|_{A_0}^{1-\theta} \, \|f\|_{A_1}^{\theta} \tag{7.3}$$

holds for some constant C and all $f \in \Delta(\bar{A})$.

To see, for instance, that (7.2) implies (7.3), we note that with

$$t = \frac{\|f\|_{A_0}}{\|f\|_{A_1}}$$

we have

$$\begin{aligned}\|f\|_X &\le Ct^{-\theta}J(t,f;\bar{A})\\ &= C\left(\frac{\|f\|_{A_0}}{\|f\|_{A_1}}\right)^{-\theta}\max\left\{\|f\|_{A_0}, \frac{\|f\|_{A_0}}{\|f\|_{A_1}}\|f\|_{A_1}\right\}\\ &= C\|f\|_{A_0}^{1-\theta}\|f\|_{A_1}^{\theta}.\end{aligned}$$

Suppose, conversely, that (7.3) holds, and let $t > 0$. Then

$$\begin{aligned}\|f\|_X &\le Ct^{-\theta}\|f\|_{A_0}^{1-\theta}(t\|f\|_{A_1})^{\theta}\\ &\le Ct^{-\theta}(\max\{\|f\|_{A_0}, t\|f\|_{A_1})^{1-\theta}(\max\{\|f\|_{A_0}, t\|f\|_{A_1})^{\theta}\\ &= Ct^{-\theta}J(t,f;\bar{A}).\end{aligned}$$

An intermediate space X is said to belong to the class $\mathcal{C}_K = \mathcal{C}_K(\theta;\bar{A})$ if for some constant C it holds that

$$K(t,f;\bar{A}) \le Ct^{\theta}\|f\|_X$$

for all $f \in X$.

7.2.5 *Reiteration*

The real interpolation method possesses a remarkable and useful stability property, which is described in the following result.

Proposition 7.5 (The Reiteration Theorem) Consider the two couples $\bar{A} = (A_0, A_1)$ and $\bar{X} = (X_0, X_1)$ of Banach spaces. Suppose that $0 < \theta_0 < \theta_1 < 1$, and put, for some $\eta \in (0,1)$,

$$\theta = (1-\eta)\theta_0 + \eta\theta_1.$$

(a) If X_i is of class $\mathcal{C}_J(\theta_i;\bar{A})$, $i = 0, 1$, then, for any $q \in [1,\infty]$,

$$\bar{A}_{\theta,q} \subseteq \bar{X}_{\eta,q}.$$

(b) If X_i is of class $\mathcal{C}_K(\theta_i;\bar{A})$, $i = 0, 1$, then, for any $q \in [1,\infty]$,

$$\bar{A}_{\theta,q} \supseteq \bar{X}_{\eta,q}.$$

7.2.6 *Interpolation of Weighted Lebesgue Spaces*

In order to connect Carlson type inequalities with interpolation theory, we need to identify the real interpolation spaces produced from a couple of weighted Lebesgue spaces. We will only need the following special case of this type of results.

Proposition 7.6 Let

$$A_i = L_{r_i}(v_i^{r_i}\,\mu), \quad i = 0, 1$$

and suppose that

$$\frac{1}{r} = \frac{1-\theta}{r_0} + \frac{\theta}{r_1}.$$

Then

$$\bar{A}_{r,\theta} = L_r(v^r\,\mu),$$

where

$$v = v_0^{1-\theta} v_1^{\theta}.$$

7.3 Embeddings of Real Interpolation Spaces

In general, the scale $\bar{A}_{\theta,q}$ of real interpolation spaces is increasing with q. Thus, in particular,

$$\bar{A}_{\theta,1} \subseteq \bar{A}_{\theta,p}$$

for any $p \geq 1$. Suppose now that

$$X = L_p(w^p\,\mu)$$

and

$$A_i = L_{p_i}(w_i^{p_i}\,\mu), \quad i = 0, 1.$$

Theorem 6.1 then gives conditions on the weights w and w_i in order for the inequality (7.3) to hold. Moreover, according to Proposition 7.4, the same conditions also give the embedding

$$\bar{A}_{\theta,1} \subseteq X$$

of the real interpolation space $\bar{A}_{\theta,1}$ into X. By combining Theorem 6.1 with the results from interpolation theory presented in this chapter, we can, in fact, prove the following stronger theorem by L. Larsson [51].

Theorem 7.1 Suppose that the hypotheses of Theorem 6.1 hold. Then

$$\bar{A}_{\theta,p} \subseteq X.$$

Proof. Take θ_i, $i = 0, 1$, such that $0 < \theta_0 < \theta < \theta_1 < 1$, and define

$$\eta = \frac{\theta - \theta_0}{\theta_1 - \theta_0}.$$

For $i = 0, 1$, let

$$v_i = w\left(\frac{w_0}{w_1}\right)^{\theta - \theta_i}$$

and define r_i by

$$\frac{1}{r_i} = \frac{1}{q} + \frac{1 - \theta_i}{p_0} + \frac{\theta_i}{p_1}.$$

Let $X_i = L_{r_i}(v_i^{r_i}\,\mu)$, $i = 0, 1$. By assumption,

$$\frac{1}{r_i} - \frac{1 - \theta_i}{p_0} - \frac{\theta_i}{p_1} = \frac{1}{q} \geq 0$$

and

$$\frac{v_i}{w_0^{1-\theta_i} w_1^{\theta_i}} = \frac{w}{w_0^{1-\theta} w_1^{\theta}} \in L_s(\mu),$$

and hence it follows from Theorem 6.1 that there are constants C_i such that

$$\|f\|_{X_i} \leq C_i \, \|f\|_{A_0}^{1-\theta_i} \, \|f\|_{A_1}^{\theta_i}, \quad i = 0, 1.$$

In other words, $X_i \in \mathcal{C}_J(\theta_i; \bar{A})$, $i = 0, 1$. By using part (a) of the Reiteration Theorem (Proposition 7.5), it follows that

$$\bar{A}_{\theta,p} \subseteq \bar{X}_{\eta,p}.$$

Now,

$$v_0^{1-\eta} v_1^{\eta} = w$$

and

$$\frac{1-\eta}{r_0} + \frac{\eta}{r_1} = \frac{1}{p},$$

so Proposition 7.6 implies that

$$\bar{X}_{\eta,p} = X.$$

This completes the proof. $\square$

Since the scale $\bar{A}_{\theta,r}$ increases with r, the conclusion of Theorem 7.1 holds true if p is replaced by any $r \in [1, p)$. However, the following partial converse shows that we cannot, in general, go beyond p.

Proposition 7.7 Given $p, p_0, p_1 \in (0, \infty]$ and $\theta \in (0,1)$ satisfying (6.1), for any $r \in (p, \infty]$, there is a measure space (Ω, μ) and weights w, w_0 and w_1 satisfying (6.2) and (6.3) such that

$$\bar{A}_{\theta,r} \not\subseteq X.$$

Proof. Let $\Omega = (e, \infty)$ and define the measure μ by

$$d\mu(x) = \frac{dx}{x},$$

where dx, as usual, denotes Lebesgue measure. Define $w(x) = 1$, $w_0(x) = x^{\theta}$, and $w_1(x) = x^{-(1-\theta)}$. Then

$$\frac{w(x)}{w_0^{1-\theta}(x) w_1^{\theta}(x)} = 1, \quad x \in \Omega,$$

so the condition (6.2) is satisfied with $s = \infty$. Moreover, since

$$\frac{w_0(x)}{w_1(x)} = x,$$

we have

$$\begin{aligned} \mu\left(\left\{2^m \leq \frac{w_0}{w_1} < 2^{m+1}\right\}\right) &= \int_{2^m}^{2^{m+1}} \frac{dx}{x} \\ &= \log 2^{m+1} - \log 2^m = \log 2, \end{aligned}$$

so that (6.3) holds with $B = \log 2$. However, consider the function

$$f(x) = \frac{1}{(\log x)^{1/p}}.$$

If $0 < t < e$, define $f_0^{(t)} = 0$ and $f_1^{(t)} = f$. Then $\left\| f_0^{(t)} \right\|_{A_0} = 0$, while

$$\left\| f_1^{(t)} \right\|_{A_1} = \left(\int_e^\infty \frac{x^{-p_1(1-\theta)}}{(\log x)^{p_1/p}} \frac{dx}{x} \right)^{1/p_1} = D < \infty,$$

and hence

$$\begin{aligned} t^{-\theta} K(t, f; \bar{A}) &\leq t^{-\theta} \left\| f_0^{(t)} \right\|_{A_0} + t^{1-\theta} \left\| f_1^{(t)} \right\|_{A_1} \\ &= D t^{1-\theta}, \quad 0 < t < e. \end{aligned}$$

If $t > e$, put $f_0^{(t)} = f\chi_{(e,t)}$ and $f_1^{(t)} = f\chi_{[t,\infty)}$. Then $f_0^{(t)} + f_1^{(t)} = f$ for all t, and

$$\begin{aligned} \int_\Omega (f_0^{(t)} w_0)^{p_0}\, d\mu &= \int_e^t \frac{x^{\theta p_0}}{(\log x)^{p_0/p}} \frac{dx}{x} \\ &\leq D_0^{p_0} \frac{t^{\theta p_0}}{(\log t)^{p_0/p}} \end{aligned}$$

and similarly

$$\int_\Omega (f_1^{(t)} w_1)^{p_1}\, d\mu \leq D_1^{p_1} \frac{t^{-(1-\theta)p_1}}{(\log t)^{p_1/p}}.$$

Thus

$$t^{-\theta} \left\| f_0^{(t)} \right\|_{A_0} + t^{1-\theta} \left\| f_1^{(t)} \right\|_{A_1} \leq \frac{D_0 + D_1}{(\log t)^{1/p}} \quad t > e.$$

It follows that

$$\|f\|_{\theta,r}^r \leq D^r \int_0^e t^{(1-\theta)r} \frac{dt}{t} + (D_0 + D_1)^r \int_e^\infty \frac{dt}{t(\log t)^{r/p}}.$$

Since $r > p$, the last integral converges (and so does the first). We see that $f \in \bar{A}_{\theta,r}$. However, since

$$\int_\Omega (fw)^p\, d\mu = \int_e^\infty \frac{dx}{x \log x} = \infty,$$

it holds that $f \notin X$. Thus

$$\bar{A}_{\theta,r} \not\subseteq X,$$

as claimed. $\square$

Remark 7.1 Although we will refrain from doing so here, it is possible to prove a multi-dimensional version of Theorem 7.1, using the corresponding multi-dimensional Theorem 6.4 (see [51]).

Remark 7.2 We mention that the interpolation spaces we get from applying the real method to a couple of weighted Lebesgue spaces (in off-diagonal cases, i.e. when the relation

$$\frac{1}{r} = \frac{1-\theta}{r_0} + \frac{\theta}{r_1}$$

in Proposition 7.6 fails) have been characterized by several authors, see e.g. L. Maligranda and L.-E. Persson [64]. By using such results we can obtain further refinements of some of the results in this chapter.

Remark 7.3 (More general embeddings of interpolation spaces) The proof of Theorem 7.1 relies on the fact that we have a Carlson type inequality associated to the spaces involved, and that we can rewrite the space X as a real interpolation space and apply the Reiteration Theorem. We can state the conclusion in more general terms, namely, as soon as we can prove a Carlson type inequality for the auxiliary spaces X_0 and X_1, or equivalently, show that $X_i \in \mathcal{C}_J(\theta_i; \bar{A})$, $i = 0, 1$, we have the embedding

$$\bar{A}_{\theta,p} \subseteq \bar{X}_{\eta,p}.$$

The spaces X and A_i, $i = 0, 1$, need not necessarily be weighted L_p spaces, as long as there is a Carlson type inequality and a corresponding reiteration theorem. Much more general embeddings could be achieved by using the method used for the proof of Theorem 7.1.

Chapter 8

Further Connection to Interpolation Theory, the Peetre $\langle\cdot\rangle_\varphi$ Method

8.1 Introduction

In order to prepare for the discussion that will follow, we present an inequality which, in a sense that will become clear below, is equivalent to Carlson's inequality.

Proposition 8.1 The following two statements are equivalent.

(A) There is a constant C_1 such that for all sequences $\{a_k\}_{k=1}^{\infty}$ of non-negative numbers, the inequality

$$\left(\sum_{k=1}^{\infty} a_k\right)^4 \leq C_1 \sum_{k=1}^{\infty} a_k^2 \sum_{k=1}^{\infty} k^2 a_k^2 \tag{8.1}$$

holds.

(B) There is a constant C_2 such that for all sequences $\{b_l\}_{l=0}^{\infty}$ of non-negative numbers, the inequality

$$\left(\sum_{l=0}^{\infty} b_l\right)^4 \leq C_2 \sum_{l=0}^{\infty} \frac{b_l^2}{2^l} \sum_{l=0}^{\infty} 2^l b_l^2 \tag{8.2}$$

holds.

Proof. Suppose that (8.1) holds. For $l \geq 0$, put

$$a_k = \frac{b_l}{2^l}, \quad 2^l \leq k < 2^{l+1}.$$

Then

$$\sum_{k=1}^{\infty} a_k = \sum_{l=0}^{\infty} \sum_{k=2^l}^{2^{l+1}-1} \frac{b_l}{2^l} = \sum_{l=0}^{\infty} \frac{b_l}{2^l}(2^{l+1} - 2^l) = \sum_{l=0}^{\infty} b_l$$

and

$$\sum_{k=1}^{\infty} a_k^2 = \sum_{l=1}^{\infty} \sum_{k=2^l}^{2^{l+1}-1} \left(\frac{b_l}{2^l}\right)^2 = \sum_{l=0}^{\infty} \frac{b_l^2}{2^{2l}}(2^{l+1} - 2^l) = \sum_{l=0}^{\infty} \frac{b_l^2}{2^l}.$$

Moreover,

$$\sum_{k=2^l}^{2^{l+1}-1} k^2 \leq 2 \cdot 2^{3l}$$

for all l, and therefore

$$\begin{aligned} \sum_{k=1}^{\infty} k^2 a_k^2 &= \sum_{l=0}^{\infty} \sum_{k=2^l}^{2^{l+1}-1} k^2 \left(\frac{b_l}{2^l}\right)^2 = \sum_{l=0}^{\infty} \frac{b_l^2}{2^{2l}} \sum_{k=2^l}^{2^{l+1}-1} k^2 \\ &\leq \sum_{l=0}^{\infty} \frac{b_l^2}{2^{2l}} 2 \cdot 2^{3l} = 2 \sum_{l=0}^{\infty} 2^l b_l^2. \end{aligned}$$

Thus by (8.1)

$$\begin{aligned} \left(\sum_{l=0}^{\infty} b_l\right)^4 &= \left(\sum_{k=1}^{\infty} a_k\right)^4 < C_1 \sum_{k=1}^{\infty} a_k^2 \sum_{k=1}^{\infty} k^2 a_k^2 \\ &\leq C_1 \sum_{l=0}^{\infty} \frac{b_l^2}{2^l} \cdot 2 \sum_{l=0}^{\infty} 2^l b_l^2 = C_2 \sum_{l=0}^{\infty} \frac{b_l^2}{2^l} \sum_{l=0}^{\infty} 2^l b_l^2, \end{aligned}$$

where $C_2 = 2C_1$, and this is (8.2).

Suppose, conversely, that (8.2) holds. Put

$$b_l = \left(\sum_{k=2^l}^{2^{l+1}-1} k a_k^2\right)^{1/2}, \quad l = 0, 1, \ldots.$$

We note that by the Schwarz inequality

$$\begin{aligned}\sum_{k=2^l}^{2^{l+1}-1} a_k &= \sum_{k=2^l}^{2^{l+1}-1} \frac{1}{k^{1/2}} k^{1/2} a_k \\ &\leq \left(\sum_{k=2^l}^{2^{l+1}-1} \frac{1}{k}\right)^{1/2} \left(\sum_{k=2^1}^{2^{l+1}-1} k a_k^2\right)^{1/2} \leq b_l,\end{aligned}$$

because

$$\sum_{k=2^l}^{2^{l+1}-1} \frac{1}{k} \leq 1$$

for all $l \geq 0$. Furthermore, since

$$\begin{aligned}\sum_{l=0}^{\infty} \frac{b_l^2}{2^l} &= \sum_{l=0}^{\infty} \frac{1}{2^l} \sum_{k=2^l}^{2^{l+1}-1} k a_k^2 \\ &\leq \sum_{l=0}^{\infty} \sum_{k=2^l}^{2^{l+1}-1} \frac{2k}{k+1} a_k^2 \\ &\leq 2 \sum_{l=0}^{\infty} \sum_{k=2^l}^{2^{l+1}-1} a_k^2 = 2 \sum_{k=1}^{\infty} a_k^2\end{aligned}$$

and similarly

$$\sum_{l=0}^{\infty} 2^l b_l^2 \leq \sum_{k=1}^{\infty} k^2 a_k^2,$$

it follows from (8.2) that

$$\begin{aligned}\left(\sum_{k=1}^{\infty} a_k\right)^4 &\leq \left(\sum_{l=0}^{\infty} b_l\right)^4 \\ &\leq C_2 \sum_{l=0}^{\infty} \frac{b_l^2}{2^l} \sum_{l=0}^{\infty} 2^l b_l^2 \\ &\leq 2C_2 \sum_{k=1}^{\infty} a_k^2 \sum_{k=1}^{\infty} k^2 a_k^2,\end{aligned}$$

which is (8.1) with $C_1 = 2C_2$. □

8.2 Carlson Type Inequalities as Sharpenings of Jensen's Inequality

Let $\varphi : \mathbb{R}_+ \to \mathbb{R}_+$ be any concave function. Then φ is necessarily non-decreasing. Indeed, the concavity of φ implies that if $0 < t_1 < t_2 < \infty$, then

$$\varphi(t) \leq \frac{\varphi(t_2) - \varphi(t_1)}{t_2 - t_1}(t - t_1) + \varphi(t_1) \tag{8.3}$$

whenever $t < t_1$ or $t > t_2$. Assume that there are $t_1 < t_2$ such that $\varphi(t_1) > \varphi(t_2)$, say

$$\varphi(t_2) - \varphi(t_1) = -\delta(t_2 - t_1)$$

for some $\delta > 0$. Then by (8.3) we have

$$\varphi(t) \leq -\delta(t - t_1) + \varphi(t_1) < 0$$

for t sufficiently large. This is a contradiction, and hence φ is non-decreasing. Moreover, the function

$$t \mapsto \frac{\varphi(t)}{t}$$

is non-increasing. Assume, namely, that there are $t_1 < t_2$ such that

$$\frac{\varphi(t_1)}{t_1} < \frac{\varphi(t_2)}{t_2},$$

say

$$t_1\varphi(t_2) - t_2\varphi(t_1) = \delta[\varphi(t_2) - \varphi(t_1)],$$

where $\delta > 0$. By (8.3)

$$\varphi(t) \leq \frac{\varphi(t_2) - \varphi(t_1)}{t_2 - t_1}t - \frac{\delta}{t_2 - t_1} < 0$$

if t is sufficiently small. This contradiction shows that $t \mapsto \varphi(t)/t$ is non-increasing.

Define the function ψ of two variables by

$$\psi(s,t) = \begin{cases} s\varphi\left(\frac{t}{s}\right), & s,t > 0, \\ 0, & s = 0 \text{ or } t = 0. \end{cases} \tag{8.4}$$

Note that ψ so defined is non-decreasing in each variable separately. This can be seen by an argument similar to those used above to show the monotonicity statements about φ.

With $\varphi(t) = t^{\frac{1}{2}}$, the inequality (8.2) can be written as

$$\sum_{l=0}^{\infty} b_l \leq C_2^{\frac{1}{4}} \psi \left(\left(\sum_{l=0}^{\infty} \frac{b_l^2}{\varphi(2^l)^2} \right)^{\frac{1}{2}}, \left(\sum_{l=0}^{\infty} 2^{2l} \frac{b_l^2}{\varphi(2^l)^2} \right)^{\frac{1}{2}} \right)$$

or

$$\|\{b_l\}_l\|_{l_1} \leq C_2^{\frac{1}{4}} \psi \left(\left\| \left\{ \frac{b_l}{\varphi(2^l)} \right\}_l \right\|_{l_2}, \left\| \left\{ 2^l \frac{b_l}{\varphi(2^l)} \right\}_l \right\|_{l_2} \right).$$

We can consider this inequality for any concave function φ, and the l_2-norms can be replaced by l_r-norms for any r. The inequality under consideration is thus

$$\|\{b_l\}_l\|_{l_1} \leq C \psi \left(\left\| \left\{ \frac{b_l}{\varphi(2^l)} \right\}_l \right\|_{l_p}, \left\| \left\{ 2^l \frac{b_l}{\varphi(2^l)} \right\}_l \right\|_{l_q} \right). \tag{8.5}$$

We will return to this inequality shortly.

Let $\mathcal{P}$ denote the class of concave functions $\varphi : \mathbb{R}_+ \to \mathbb{R}_+$ such that

$$\lim_{t \to 0^0} \varphi(t) = 0 \quad \text{and} \quad \lim_{t \to \infty} \varphi(t) = \infty.$$

We also define the following subclasses of $\mathcal{P}$. Define

$$s_\varphi(t) = \sup_{s>0} \frac{\varphi(st)}{\varphi(s)}, \quad t > 0. \tag{8.6}$$

- $\mathcal{P}_+$ is the set of concave functions φ for which

$$\lim_{t \to 0^+} s_\varphi(t) = 0.$$

- $\mathcal{P}_-$ is the set of convave functions φ for which

$$\lim_{t \to \infty} \frac{s_\varphi(t)}{t} = 0.$$

- $\mathcal{P}_\pm = \mathcal{P}_+ \cap \mathcal{P}_-$.
- $\mathcal{P}_0$ is the set of concave functions φ satisfying

$$\lim_{t \to 0^+} \varphi(t) = 0 \quad \text{and} \quad \lim_{t \to \infty} \frac{\varphi(t)}{t} = 0.$$

Example 8.1 Let

$$\varphi(t) = \frac{t}{1+t}.$$

Then, as the reader may check, we have

$$s_\varphi(t) = \max\{1, t\}.$$

Thus this gives an example of a function which is in $\mathcal{P}_0$ but not in $\mathcal{P}_\pm$.

Suppose that $\varphi \in \mathcal{P}$, so that $\psi : \mathbb{R}_+^2 \to \mathbb{R}_+$ as defined by (8.4) is concave, positively homogeneous of degree 1 with $\psi(0,0) = 0$. Then *Jensen's inequality*

$$\sum_{k=1}^{m} \psi(x_k, y_k) \le \psi\Big(\sum_{k=1}^{m} x_k, \sum_{k=1}^{m} y_k\Big) \tag{8.7}$$

holds for all (finite) sequences $x_1, \ldots, x_m$, $y_1, \ldots, y_m$ of non-negative numbers (see A.-P. Calderón [19], p. 162). In particular, for

$$\psi(u, v) = u^{\frac{1}{p}} v^{\frac{1}{p'}},$$

with $1 < p < \infty$ and

$$\frac{1}{p} + \frac{1}{p'} = 1,$$

we have the classical Hölder–Rogers inequality

$$\sum_{k=1}^{m} x_k^{\frac{1}{p}} y_k^{\frac{1}{p'}} \le \Big(\sum_{k=1}^{m} x_k\Big)^{\frac{1}{p}} \Big(\sum_{k=1}^{m} y_k\Big)^{\frac{1}{p'}}.$$

We want to prove an estimate stronger than (8.7), namely

$$\sum_{k=1}^{m} \psi(x_k, y_k) \le C\psi\Big(\big\|\{x_k\}_{k=1}^m\big\|_{l_p}, \big\|\{y_k\}_{k=1}^m\big\|_{l_q}\Big), \tag{8.8}$$

with $p, q > 1$ and C independent of m. The inequality (8.8) is not true in general. In fact, if $1 < p \le q < \infty$ and $a > 0$, put $x_k = y_k = a$, $k = 1, \ldots, m$. Then (8.8) means

$$\begin{aligned} m\psi(a,a) &\le C\psi\big(am^{\frac{1}{p}}, am^{\frac{1}{q}}\big) \\ &= Cm^{\frac{1}{p}}\psi\big(a, am^{\frac{1}{q}-\frac{1}{p}}\big) \\ &\le Cm^{\frac{1}{p}}\psi(a,a). \end{aligned}$$

This shows that C must depend m. Also, if $\lambda > 0$ and $x_k = \lambda y_k$, $k = 1, \ldots, m$, then (8.8) means

$$\begin{aligned}
\sum_{k=1}^{m} y_k \psi(\lambda, 1) &= \sum_{k=1}^{m} \psi(\lambda y_k, y_k) \\
&\le C\psi\left(\lambda\Big(\sum_{k=1}^{m} y_k^p\Big)^{\frac{1}{p}}, \Big(\sum_{k=1}^{m} y_k^q\Big)^{\frac{1}{q}}\right) \\
&= C\Big(\sum_{k=1}^{m} y_k^p\Big)^{\frac{1}{p}} \psi\left(\lambda, \Big(\sum_{k=1}^{m} y_k^p\Big)^{\frac{1}{p}} \Big(\sum_{k=1}^{m} y_k^q\Big)^{-\frac{1}{q}}\right) \\
&\le C\Big(\sum_{k=1}^{m} y_k^p\Big)^{\frac{1}{p}} \psi(\lambda, 1),
\end{aligned}$$

which is impossible with C independent of m.

For

$$x_k = \frac{a_k}{\varphi(2^k)}, \quad y_k = \frac{2^k a_k}{\varphi(2^k)},$$

where $\varphi(2^k) = \psi(2^k, 1)$, the inequality (8.8) becomes

$$\sum_{k=1}^{m} a_k \le C\psi\left(\left\|\left\{\frac{a_k}{\varphi(2^k)}\right\}\right\|_{l_p}, \left\|\left\{\frac{2^k a_k}{\varphi(2^k)}\right\}\right\|_{l_q}\right). \tag{8.9}$$

The question here is: for which p, q and ψ does there exist a constant C independent of m such that the inequality (8.9) holds? The answer was given by J. Gustavsson and J. Peetre [31] (see also [61], pp. 143–145). The inequality (8.9) holds in the following four cases:

1° $p = q = 1$ and $\varphi \in \mathcal{P}$.
2° $p = 1$, $q > 1$ and $\varphi \in \mathcal{P}_-$.
3° $p > 1$, $q = 1$ and $\varphi \in \mathcal{P}_+$.
4° $p > 1$, $q > 1$ and $\varphi \in \mathcal{P}_\pm$.

Theorem 8.1 (Gustavsson–Peetre, 1977) Suppose that $1 < p, q \le \infty$. Then, if $\varphi \in \mathcal{P}_\pm$, there is a constant C such that (8.9) holds for all sequences $\{a_k\}$.

Remark 8.1 If we take $\varphi(t) = \min\{1, t\}$, then the inequality (8.9) is not true.

Proof of Theorem 8.1. For $u > 0$ we have by the Hölder–Rogers inequality

$$\begin{aligned}
\sum_{k=1}^{m} a_k &= \sum_{2^k \le u} a_k + \sum_{2^k > u} a_k \\
&\le \left(\sum_{2^k \le u} \varphi(2^k)^{p'} \right)^{\frac{1}{p'}} \left\| \left\{ \frac{a_k}{\varphi(2^k)} \right\}_{k=1}^{m} \right\|_{l_p} \\
&+ \left(\sum_{2^k > u} \left(\frac{\varphi(2^k)}{2^k} \right)^{q'} \right)^{\frac{1}{q'}} \left\| \left\{ \frac{2^k a_k}{\varphi(2^k)} \right\}_{k=1}^{m} \right\|_{l_q} \\
&\le 4 \left(\int_0^u \varphi(t)^{p'} \frac{dt}{t} \right)^{\frac{1}{p'}} \left\| \left\{ \frac{a_k}{\varphi(2^k)} \right\}_{k=1}^{m} \right\|_{l_p} \\
&+ 4 \left(\int_u^\infty \left(\frac{\varphi(t)}{t} \right)^{q'} \frac{dt}{t} \right)^{\frac{1}{q'}} \left\| \left\{ \frac{2^k a_k}{\varphi(2^k)} \right\}_{k=1}^{m} \right\|_{l_q} \\
&\le 4 D_\varphi^0 \varphi(u) \left\| \left\{ \frac{a_k}{\varphi(2^k)} \right\}_{k=1}^{m} \right\|_{l_p} + 4 D_\varphi^1 \frac{\varphi(u)}{u} \left\| \left\{ \frac{2^k a_k}{\varphi(2^k)} \right\}_{k=1}^{m} \right\|_{l_q}
\end{aligned}$$

where D_φ^i, $i = 0, 1$ are constants depending on φ. If we put

$$u = \left\| \left\{ \frac{2^k a_k}{\varphi(2^k)} \right\}_{k=1}^{m} \right\|_{l_q} \Big/ \left\| \left\{ \frac{a_k}{\varphi(2^k)} \right\}_{k=1}^{m} \right\|_{l_p},$$

we obtain (8.9) with C depending on p, q and φ. □

Remark 8.2 If $\varphi(t) = t^\theta$, $0 < \theta < 1$ and $p > 1$, $q > 1$, then the constant $C = C(\theta, p, q)$ increases to ∞ when $\theta \to 0^+$ or $\theta \to 1^-$. To see this, we note that the inequality in this case means

$$\sum_{k=1}^{m} a_k \le C \left\| \left\{ \frac{a_k}{\varphi(2^k)} \right\}_{k=1}^{m} \right\|_{l_p}^{1-\theta} \left\| \left\{ \frac{2^k a_k}{\varphi(2^k)} \right\}_{k=1}^{m} \right\|_{l_q}^{\theta}. \tag{8.10}$$

If we take $a_k = 1$, $k = 1, \ldots, m$, (8.10) becomes

$$m \le C \left(\sum_{k=1}^{m} 2^{-\theta k p} \right)^{\frac{1-\theta}{p}} \left(\sum_{k=1}^{m} 2^{(1-\theta)kq} \right)^{\frac{\theta}{q}}$$

or

$$m \le C \left(\frac{2^{\theta pm} - 1}{2^{\theta pm}(2^{\theta p} - 1)} \right)^{\frac{1-\theta}{p}} \left(2^{(1-\theta)q} \frac{2^{(1-\theta)qm} - 1}{2^{(1-\theta)q} - 1} \right)^{\frac{\theta}{q}}.$$

When $\theta \to 0^+$, the last estimate means $m \le Cm^{\frac{1}{p}}$, which is impossible for large m. When $\theta \to 1^-$, the same estimate means $m \le Cm^{\frac{1}{q}}$, which is again impossible. Thus the constant depends on θ; it has the order

$$\theta^{-\frac{1}{p'}} + (1-\theta)^{-\frac{1}{q'}}.$$

This observation shows that [78], Lemma 4.6, for $q = 2$ with $C = 1$ is false. The author there referred to Gustavsson–Peetre [31], where they proved the estimate (8.9) with C depending on φ.

8.3 The Peetre Interpolation Method and Interpolation of Orlicz Spaces

Let $\bar{A} = (A_0, A_1)$ be a compatible couple of (quasi-)Banach spaces, and suppose that $\varphi : \mathbb{R}_+ \to \mathbb{R}_+$ is (equivalent to) a concave function. Then $\langle \bar{A} \rangle_\varphi$ is defined as the space of $a \in \Sigma(\bar{A})$ for which there is a sequence $\{u_l\}_{l \in \mathbb{Z}}$ of elements in $\Delta(\bar{A})$ such that

(*i*) $a = \sum_{l=-\infty}^{\infty} u_l$ (in $\Sigma(\bar{A})$) and

(*ii*) for some constant C it holds that

$$\left\| \sum_l \xi_l \frac{u_l}{\varphi(2^l)} \right\|_{A_0} \le C$$

and

$$\left\| \sum_l \xi_l 2^l \frac{u_l}{\varphi(2^l)} \right\|_{A_1} \le C$$

for all *finite* sequences $\{\xi_l\}_{l \in F}$ with $|\xi_l| \le 1$. Here, F is some finite subset of $\mathbb{Z}$.

On $\langle \bar{A} \rangle_\varphi$, we have the semi(quasi-)norm

$$\|a\|_{\langle \bar{A} \rangle_\varphi} = \inf C,$$

where the infimum is taken over all admissible C in (*ii*).

Remark 8.3 The construction of the space $\langle \bar{A} \rangle_\varphi$ is often referred to as the $\pm$ (interpolation) method.

This method has the following interpolation property (cf. Maligranda [61], p. 133).

Theorem 8.2 Let $\bar{A} = (A_0, A_1)$ and $\bar{B} = (B_0, B_1)$ be any couples of (quasi-)Banach spaces, and $\varphi \in \mathcal{P}$. Moreover, let $T : \bar{A} \to \bar{B}$ be any continuous linear mapping, meaning that T maps $\Sigma(\bar{A})$ into $\Sigma(\bar{B})$ and $T|_{A_i} : A_i \to B_i$, $i = 0, 1$ (recall that this statement means that the mapping is continuous). Then

$$T : \langle \bar{A} \rangle_\varphi \to \langle \bar{B} \rangle_\varphi,$$

and

$$\|T\|_{\langle \bar{A} \rangle_\varphi \to \langle \bar{B} \rangle_\varphi} \leq \max\left\{ \|T\|_{A_0 \to B_0}, \|T\|_{A_1 \to B_1} \right\}.$$

In 1977, Gustavsson–Peetre [31] used their Theorem 8.1 as a step in the proof of the following embedding result, where the Peetre $\pm$ method is applied to couples of Orlicz spaces. Let (Ω, μ) be a measure space, and let $\Phi : [0, \infty) \to [0, \infty)$ be an *Orlicz function*, i.e. an increasing, continuous, convex function. Then the *Orlicz space* $L^\Phi = L^\Phi(\Omega, \mu)$ is the spaces of (equivalence classes of) functions

$$f : \Omega \to \mathbb{R} \text{ or } \mathbb{C}$$

such that for some $\lambda > 0$ (depending on f) it holds that

$$\int_\Omega \Phi(\lambda |f(\omega)|) \, d\mu(\omega) < \infty.$$

On L^Φ, we introduce the *Luxemburg–Nakano norm* given by

$$\|f\|_\Phi = \inf\left\{ \lambda > 0; \int \Phi\left(\frac{|f(\omega)|}{\lambda} \right) d\mu(\omega) \leq 1 \right\}.$$

Theorem 8.3 (Gustavsson–Peetre, 1977) Let $\varphi \in \mathcal{P}_\pm$, and suppose that the functions Φ, Φ_0 and Φ_1 are related by the formula

$$\Phi^{-1} = \Phi_0^{-1} \varphi\left(\frac{\Phi_1^{-1}}{\Phi_0^{-1}} \right). \tag{8.11}$$

Then

$$\langle L^{\Phi_0}, L^{\Phi_1} \rangle_\varphi = L^\Phi \tag{8.12}$$

with equivalent norms.

Remark 8.4 If $\varphi \in \mathcal{P}_0$, then we can prove the embedding

$$L^\Phi \subseteq \langle L^{\Phi_0}, L^{\Phi_1} \rangle_\varphi$$

and the norm of this embedding does not exceed 2 (cf. Maligranda [61], Lemma 14.4). The reverse embedding for $\varphi \in \mathcal{P}_\pm$ has norm depending on φ (cf. Maligranda [61], Lemma 14.6) and it increases to ∞ when we go with φ "to the boundaries" (cf. Remark 8.2 above).

We are now ready to state the following result concerning interpolation of Orlicz spaces.

Theorem 8.4 (Gustavsson–Peetre, 1977; Shestakov, 1981) Let Φ_0, Φ_1, Ψ_0 and Ψ_1 be Orlicz functions.

(a) If

$$T : (L^{\Phi_0}(\mu), L^{\Phi_1}(\mu)) \to (L^{\Psi_0}(\nu), L^{\Psi_1}(\nu))$$

and

$$\Phi^{-1} = \Phi_0^{-1} \varphi\left(\frac{\Phi_1^{-1}}{\Phi_0^{-1}}\right), \quad \Psi^{-1} = \Psi_0^{-1} \varphi\left(\frac{\Psi_1^{-1}}{\Psi_0^{-1}}\right),$$

where $\varphi \in \mathcal{P}_\pm$, then $T : L^\Phi(\mu) \to L^\Psi(\nu)$ and

$$\|T\|_{L^\Phi \to L^\Psi} \leq C \max\left\{ \|T\|_{L^{\Phi_i} \to L^{\Psi_i}}, i = 0, 1 \right\}$$

where the constant C depends on φ.

(b) If

$$T : (L^{\Phi_0}(\mu), L_\infty(\mu)) \to (L^{\Psi_0}(\nu), L_\infty(\nu))$$

and

$$\Phi^{-1} = \Phi_0^{-1} \varphi\left(\frac{1}{\Phi_0^{-1}}\right), \quad \Psi^{-1} = \Psi_0^{-1} \varphi\left(\frac{1}{\Psi_0^{-1}}\right)$$

where $\varphi \in \mathcal{P}_+$, then $T : L^\Phi(\mu) \to L^\Psi(\nu)$ and

$$\|T\|_{L^\Phi \to L^\Psi} \leq C \max\left\{ \|T\|_{L^{\Phi_0} \to L^{\Psi_0}}, \|T\|_{L_\infty \to L_\infty} \right\}$$

where the constant C depends on φ.

For detailed proofs of the above results, as well as a more extensive discussion on interpolation of Orlicz spaces, see the exposition by L. Maligranda [61], Chapter 14.

Remark 8.5 As will be seen in the next section, it is, in fact, necessary that the function φ belongs to the class $\mathcal{P}_{\pm}$ in order for the inequality (8.8) to hold, but not necessary for the interpolation property in Theorem 8.4.

Remark 8.6 Some other identifications of the Peetre method $\langle\cdot\rangle_{\varphi}$ for weighted L_p spaces, weighted Orlicz spaces or vector-valued spaces, which used the Carlson type inequality from Theorem 8.3, were proved by J. Gustavsson [30], D. L. Fernandez and J. B. Garcia [25], and J. B. Garcia [28]. This means that the constant in the interpolation theorem depends on $\varphi \in \mathcal{P}_{\pm}$. Fernandez–Garcia proved the following vector-valued version of the Carlson inequality. Let E be a Banach lattice, $\varphi \in \mathcal{P}_{\pm}$, and $1 \le p < \infty$. If $\{u_k\}_{k=1}^m$ is a finite sequence in E, then

$$\left\| \sum_{k=1}^{m} |u_k| \right\|_E \le C\psi\left(\left\| \left\{ \frac{|u_k|}{\varphi(2^k)} \right\} \right\|_{E(l_p)}, \left\| \left\{ \frac{2^k |u_k|}{\varphi(2^k)} \right\} \right\|_{E(l_p)} \right),$$

where ψ is defined as in (8.4) and the constant $C > 0$ is independent of $m \in \mathbb{N}$.

8.4 A Carlson Type Inequality with Blocks

N. Ya. Kruglyak, L. Maligranda and L.-E. Persson [47] used the so-called Brudnyĭ–Kruglyak construction, (cf. [18]) which, given a function φ in the class $\mathcal{P}_0$ gives a subdivision of $\mathbb{R}_+$ into certain subintervals, serving as a ground for *optimal blocks* used in an extension of the inequality (8.5). Before we introduce this subdivision, we present the following symmetric version of (8.5).

For $m = 1, 2, \ldots$, let S_m be the sector in $\mathbb{R}_+^2$ which is bounded by the lines $y = 2^m x$ and $y = 2^{m+1} x$ (see Figure 8.1).

Proposition 8.2 There is a constant C such that (8.5) holds if and only if for some constant C_0 we have

$$\sum_k \psi(a_k, b_k) \le C_0 \psi\left(\left\| \left\{ \sum_{(a_k, b_k) \in S_m} a_k \right\}_m \right\|_{l_p}, \left\| \left\{ \sum_{(a_k, b_k) \in S_m} b_k \right\}_m \right\|_{l_q} \right).$$

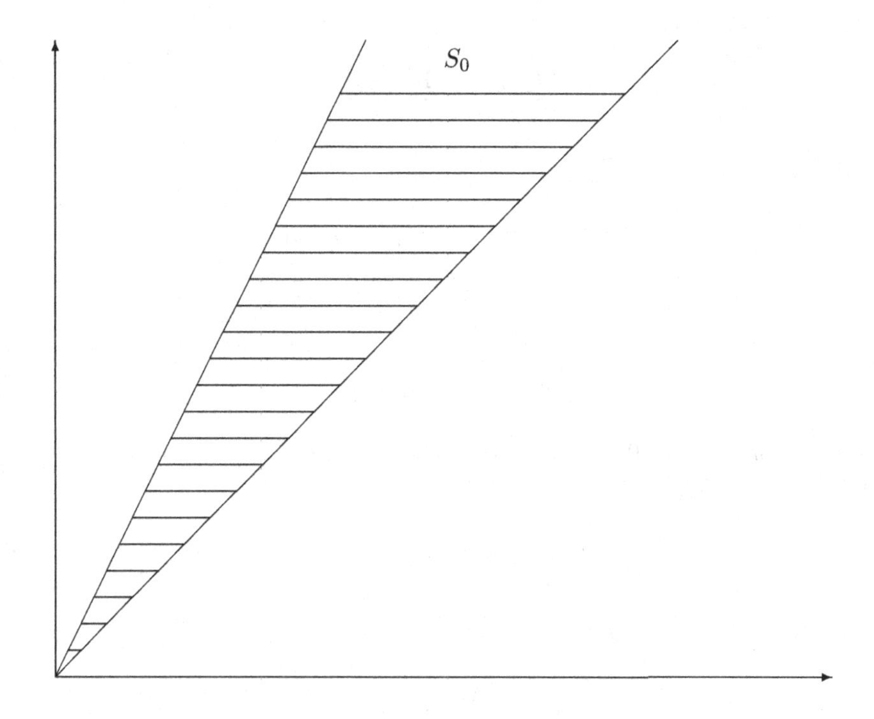

Fig. 8.1

The main result from [47], and of this section, has two parts. The first part is the important observation that in order for the inequality (8.5) to hold, it is also necessary that the function φ belongs to $\mathcal{P}_{\pm}$. The form of (8.5) thus needs to be modified if we want to cover the larger class $\mathcal{P}_0$. The second part does just this. Proposition 8.2 gives a hint of how to do it, namely, the *blocks* into which $\mathbb{R}^2_+$ is divided by the sectors S_m are modified in such a way that any function from $\mathcal{P}_0$ can be used.

Before we state the main result, we explain what the Brudnyĭ–Kruglyak construction is all about.

Lemma 8.1 (The Brudnyĭ–Kruglyak Construction) Let $\varphi \in \mathcal{P}_0$ and let $r > 1$ be fixed. Then there exists a decomposition $\{\chi_k\}$ of $\mathbb{R}_+$ into closed intervals χ_k with disjoint interiors, having the following properties, where we denote by t_{2k} the left endpoint of χ_k.

(a) For each k, we can choose $t_{2k+1} \in \chi_k$ such that

$$\frac{\varphi(t_{2k+1})}{t_{2k+1}} = \frac{1}{r}\frac{\varphi(t_{2k})}{t_{2k}}.$$

(b) If $t \in \chi_k$, then

$$\varphi(t) \leq r \min\{1, t/t_{2k+1}\}\varphi(t_{2k+1})$$

and

(c) For all k,

$$\varphi(t_{2k+2}) = r\varphi(t_{2k+1}).$$

Proof. For any $s > 0$, let χ_s be the closed subinterval of $\mathbb{R}_+$ consisting of those t for which

$$\varphi(t) \leq r\varphi(s) \min\left\{1, \frac{t}{s}\right\}$$

(see Figure 8.2). We consider now the intervals

$$\chi_k = \chi_{t_{2k+1}},$$

where the t_k are chosen so that the right endpoint t_{2k+2} of the interval χ_k coincides with the left endpoint of the interval χ_{k+1}. The properties follow from this construction. □

Remark 8.7 The construction of the sequence in Lemma 8.1 is sometimes also called the Oskolkov–Janson construction, and its inductive definition was given by the formula

$$t_0 = 1 \quad \text{and} \quad \min\left(\frac{\varphi(t_{k+1})}{\varphi(t_k)}, \frac{t_{k+1}\varphi(t_{k+1})}{t_k\varphi(t_k)}\right) = r > 1.$$

K. I. Oskolkov used this construction in the paper [69] published in 1977, S. Janson [37] in 1981, and Brudnyĭ–Kruglyak [17] in 1981.

The following two basic facts are of importance below. The first one gives an equivalent characterization of a function in the class $\mathcal{P}_0$ in terms of the Brudnyĭ–Kruglyak construction associated to it, and the second shows that the intervals χ_k are, in a sense, optimal.

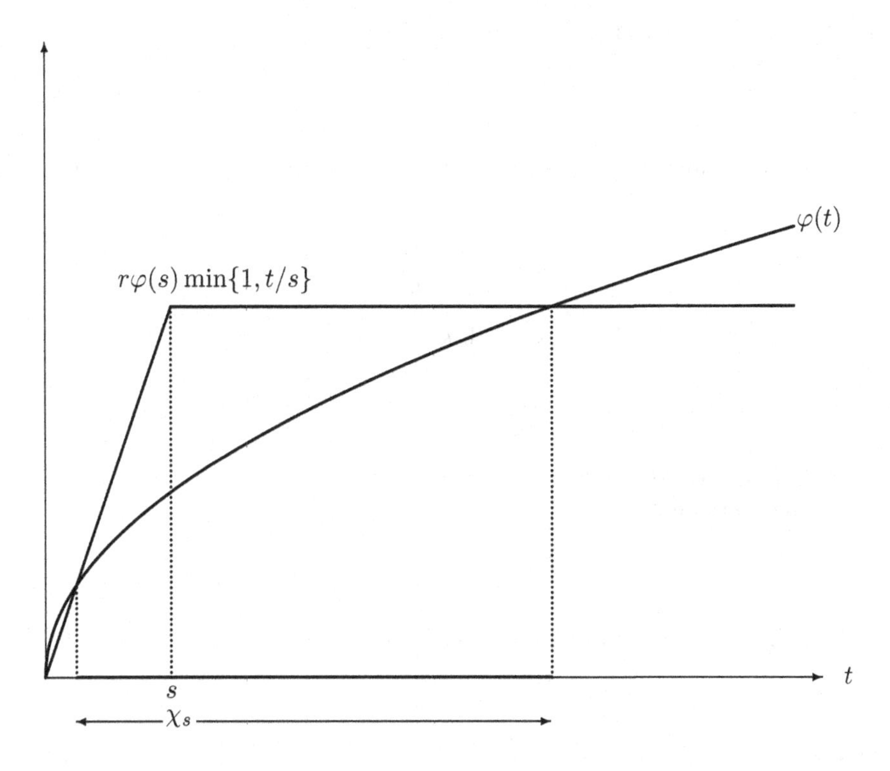

Fig. 8.2 The diagram shows how the interval χ_s is defined.

Lemma 8.2 Let $\varphi \in \mathcal{P}_0$, and let $\{t_k\}_k$ be the sequence associated to φ by the Brudnyĭ–Kruglyak construction. Then $\varphi \in \mathcal{P}_\pm$ if and only if

$$\sup_k \frac{t_{2k+2}}{t_{2k}} < \infty.$$

Proof. We assume throughout that $r > 1$ is fixed. If $\varphi \in \mathcal{P}_-$, there is $u > 1$ such that for all s

$$\frac{\varphi(su)}{\varphi(s)} < \frac{1}{r}u$$

or

$$\frac{\varphi(su)}{su} < \frac{1}{r}\frac{\varphi(s)}{s}.$$

With $s = t_{2k}$, this, together with the property (a) of Lemma 8.1 and the fact that $\varphi(t)/t$ is non-increasing, yields that

$$\frac{t_{2k+1}}{t_{2k}} < u.$$

In a similar manner, if $\varphi \in \mathcal{P}_+$, then there is v satisfying $0 < v < 1$ such that

$$\frac{t_{2k+2}}{t_{2k+1}} < \frac{1}{v}.$$

Thus, for any $\varphi \in \mathcal{P}_\pm$, it holds for all k

$$\frac{t_{2k+2}}{t_{2k}} < \frac{u}{v} < \infty,$$

so the only if part of the lemma is proved.

Assume now that

$$\frac{t_{2k+2}}{t_{2k}} \leq M < \infty$$

for all k. Let $s \in (0, \infty)$ and pick k_0 such that

$$t_{2k_0} \leq s < t_{2k_0+2}.$$

Then

$$t_{2k_0+4} \leq M t_{2k_0+2} \leq M^2 t_{2k_0} \leq M^2 s,$$

so by part (a) of Lemma 8.1, also using that the function $\varphi(t)/t$ is non-increasing, we find that

$$\frac{\varphi(M^2 s)}{M^2 s} \leq \frac{\varphi(t_{2k_0+4})}{t_{2k_0+4}} \leq \frac{\varphi(t_{2k_0+3})}{t_{2k_0+3}} = \frac{1}{r}\frac{\varphi(t_{2k_0+2})}{t_{2k_0+2}} \leq \frac{1}{r}\frac{\varphi(s)}{s}.$$

By iterating this process, we find that for any m we have

$$\varphi(M^{2m} s) \leq r^{-m} M^{2m} \varphi(s),$$

from which it follows that

$$\frac{s_\varphi(t)}{t} \to 0, \quad t \to \infty$$

(because of the continuity of φ), i.e. $\varphi \in \mathcal{P}_-$. In an analoguous way, it is shown that $\varphi \in \mathcal{P}_+$, which completes the proof of the if part. $\square$

Proposition 8.3 Suppose that $\{\Omega_k\}_{k=1}^{\infty}$ is any decomposition of the set $\{2^m\}_{m\in\mathbb{Z}}$. Then the inequality (8.13) holds for some C with the χ_k replaced by Ω_k if and only if there is a constant M such that

$$\operatorname{card}(\{m; \Omega_m \cap \chi_k \neq \emptyset\}) \leq M, \quad k \in \mathbb{Z}.$$

As a final preparation for Theorem 8.5 below, we mention that the following extended version of Proposition 8.2 holds.

Proposition 8.4 Suppose that $1 < p, q \leq \infty$. Let $\varphi \in \mathcal{P}_0$, and let $\{t_m\}_m$ be its associated Brudnyĭ–Kruglyak sequence, with the corresponding intervals $\{\chi_m\}$. Then the following two statements are equivalent.

(a) There is a constant C_1 such that

$$\sum_k a_k \leq C_1 \psi\left(\left\| \left\{ \sum_{2^k\in\chi_m} \frac{a_k}{\varphi(2^k)} \right\}_m \right\|_{l_p}, \left\| \left\{ \sum_{2^k\in\chi_m} \frac{2^k a_k}{\varphi(2^k)} \right\}_m \right\|_{l_q} \right).$$

(b) There is a constant C_2 such that

$$\sum_k \psi(a_k, b_k) \leq C_2 \psi\left(\left\| \left\{ \sum_{(a_k,b_k)\in T_m} a_k \right\}_m \right\|_{l_p}, \left\| \left\{ \sum_{(a_k,b_k)\in T_m} b_k \right\}_m \right\|_{l_q} \right).$$

As mentioned above, the main result of this section, from [47], is divided into two parts, the first of which tells us that we cannot get the inequality (8.5) unless $\varphi \in \mathcal{P}_{\pm}$, and the second of which describes how we can overcome this problem by replacing the sectors S_m by sectors arising from the Brudnyĭ–Kruglyak construction. More precisely, we replace each S_m by the sector T_m bounded by the lines $y = t_{2m}x$ and $y = t_{2m+2}x$, where the t_{2m} are the left endpoints of the intervals χ_m.

Theorem 8.5 Suppose that $1 < p, q \leq \infty$ and $\varphi \in \mathcal{P}_0$.

(a) If there is a constant C such that (8.5) holds, then $\varphi \in \mathcal{P}_{\pm}$.
(b) Let $\{\chi_m\}$ be the intervals arising from the Brudnyĭ–Kruglyak construction associated to φ. Then there is a constant C such that for any sequence $\{a_k\}$ of non-negative numbers

$$\sum a_k \leq C \psi\left(\left\| \left\{ \sum_{2^k\in\chi_m} \frac{a_k}{\varphi(2^k)} \right\}_m \right\|_{l_p}, \left\| \left\{ \sum_{2^k\in\chi_m} \frac{2^k a_k}{\varphi(2^k)} \right\}_m \right\|_{l_q} \right). \tag{8.13}$$

The best constant C in (8.13) satisfies

$$C \leq (1+\sqrt{2})^2.$$

Proof of Theorem 8.5. (a) Suppose that (8.5) holds for some C. Note that this inequality can be written as the block inequality (8.13), but with the blocks χ_k replaced by just single-point sets $\Omega_k = \{2^k\}$. Thus, if $\varphi \notin \mathcal{P}_\pm$, then Lemma 8.2 implies that for any m, we can find k such that

$$\frac{t_{2k+2}}{t_{2k}} > 2^m.$$

But this means that at least $m-1$ of the Ω_k are contained in χ_m, which contradicts Proposition 8.3. Thus $\varphi \in \mathcal{P}_\pm$.

(b) Since $\|\cdot\|_{l_\infty} \leq \|\cdot\|_{l_r}$ for any finite r and since ψ is non-decreasing in each variable, it suffices to prove (8.13) with $p = q = \infty$. We consider

$$A = \sup_m \sum_{2^k \in \chi_m} \frac{a_k}{\varphi(2^k)},$$

$$B = \sup_m \sum_{2^k \in \chi_m} \frac{2^k a_k}{\varphi(2^k)}$$

and put

$$M = \frac{B}{A}.$$

Let m_0 be the index for which $M \in \chi_{m_0}$ and make the decomposition

$$\sum a_k = \sum_{m<m_0} \sum_{2^k \in \chi_m} a_k + \sum_{2^k \in \chi_{m_0}} a_k + \sum_{m>m_0} \sum_{2^k \in \chi_m} a_k.$$

By part (b) of Lemma 8.1, it holds that

$$\max_{2^k \in \chi_m} \varphi(2^k) \leq r\varphi(t_{2m+1}).$$

Moreover, by iterating part (c) of Lemma 8.1 and using that φ is non-decreasing, we find that

$$\varphi(t_{2m_0-1}) \leq \frac{1}{r}\varphi(M),$$

$$\varphi(t_{2m_0-3}) \leq \frac{1}{r^2}\varphi(M)$$

etc. We therefore have

$$\begin{aligned}
\sum_{m<m_0} \sum_{2^k \in \chi_m} a_k &= \sum_{m<m_0} \sum_{2^k \in \chi_m} \frac{a_k}{\varphi(2^k)} \varphi(2^k) \\
&\leq \sum_{m<m_0} \sum_{2^k \in \chi_m} \frac{a_k}{\varphi(2^k)} \max_{2^k \in \chi_m} \varphi(2^k) \\
&\leq A \sum_{m<m_0} r\varphi(t_{2m+1}) \\
&\leq Ar\left(\frac{\varphi(M)}{r} + \frac{\varphi(M)}{r^2} + \dots\right) \\
&= A\varphi(M)\frac{r}{r-1} = \psi(A,B)\frac{r}{r-1}.
\end{aligned}$$

In a similar manner, we obtain

$$\sum_{m>m_0} \sum_{2^k \in \chi_m} a_k \leq \psi(A,B)\frac{r}{r-1}.$$

As for the middle term, suppose first that $\varphi(M) \geq \varphi(t_{2m_0+1})$. Then

$$\begin{aligned}
\sum_{2^k \in \chi_{m_0}} a_k &\leq \sum_{2^k \in \chi_{m_0}} \frac{a_k}{\varphi(2^k)} \max_{2^k \in \chi_{m_0}} \varphi(2^k) \\
&\leq \sum_{2^k} \frac{a_k}{\varphi(2^k)} r\varphi(t_{2m_0+1}) \\
&\leq Ar\varphi(M) = r\psi(A,B).
\end{aligned}$$

If, on the other hand, $\varphi(M) < \varphi(t_{2m_0+1})$, then

$$\begin{aligned}
\sum_{2^k \in \chi_{m_0}} a_k &\leq \sum_{2^k \in \chi_{m_0}} \frac{2^k a_k}{\varphi(2^k)} \max_{2^k \in \chi_{m_0}} \frac{\varphi(2^k)}{2^k} \\
&\leq \sum_{2^k \in \chi_{m_0}} \frac{2^k a_k}{\varphi(2^k)} r \frac{\varphi(t_{2m_0+1})}{t_{2m_0+1}} \\
&\leq Br\frac{\varphi(M)}{M} = r\psi(A,B).
\end{aligned}$$

By combining these inequalities, we obtain

$$\begin{aligned}
\sum a_k &\leq \left(\frac{r}{r-1} + r + \frac{r}{r-1}\right)\psi(A,B) \\
&= \frac{r(r+1)}{r-1}\psi(A,B).
\end{aligned}$$

The infimum over $r > 1$ is attained at $r = 1 + \sqrt{2}$ and we therefore get (8.13) with $C = (1 + \sqrt{2})^2$. The proof is complete. □

As a final remark, concluding this chapter, we note that the following extended version of Proposition 8.1 holds. It is suggested by Levin's theorem and the function $\varphi(t) = t^\theta$, where

$$\theta = \frac{q\lambda}{p\mu + q\lambda}.$$

Proposition 8.5 The following two statements are equivalent.

(A) There is a constant C_1 such that for all sequences $\{a_k\}_{k=1}^\infty$ of non-negative numbers, the inequality

$$\sum_{k=1}^\infty a_k \le C_1 \left(\sum_{k=1}^\infty k^{p-1-\lambda} a_k^p\right)^{\frac{\mu}{p\mu+q\lambda}} \left(\sum_{k=1}^\infty k^{q-1+\mu} a_k^q\right)^{\frac{\lambda}{p\mu+q\lambda}}$$

holds.

(B) There is a constant C_2 such that for all sequences $\{b_l\}_{l=0}^\infty$ of non-negative numbers, the inequality

$$\sum_{l=0}^\infty b_l \le C_2 \left(\sum_{l=0}^\infty 2^{-\frac{\lambda pq}{p\mu+q\lambda}} b_l^p\right)^{\frac{\mu}{p\mu+q\lambda}} \left(\sum_{l=0}^\infty 2^{\frac{\mu pq}{p\mu+q\lambda}} b_l^q\right)^{\frac{\lambda}{p\mu+q\lambda}}$$

or

$$\|\{b_l\}\|_{l_1} \le C_2 \psi\left(\left\|\left\{\frac{b_l}{\varphi(2^l)}\right\}_l\right\|_{l_p}, \left\|\left\{\frac{2^l b_l}{\varphi(2^l)}\right\}_l\right\|_{l_q}\right)$$

holds, where $\varphi(t) = t^{\frac{q\lambda}{p\mu+q\lambda}}$ and ψ is defined as in (8.4).

8.5 The Calderón–Lozanovskiĭ Construction on Banach Lattices

The Carlson inequality with blocks in the previous section gives a comparison of the Peetre method with the Calderón–Lozanovskiĭ construction on Banach lattices, to be described below, and the embedding constants are independent of the function φ in the construction. This gives the interpolation result with a universal constant (independent of φ) in the estimation of operator norms. Since the Calderón–Lozanovskiĭ construction for a couple of Orlicz spaces is easy to describe, we therefore also get the interpolation result for Orlicz spaces with a universal interpolation constant.

Let $\bar{X} = (X_0, X_1)$ be a Banach couple of lattices, i.e. a Banach couple of functional lattices defined on the same measure space (Ω, μ). For $\varphi \in \mathcal{P}$, we define as before $\psi(s,t) = s\varphi(t/s)$ if $s > 0$ and $t > 0$, and $\psi(s,t) = 0$ if $s = 0$ or $t = 0$. We consider the *Calderón–Lozanovskiĭ space* $\varphi(\bar{X})$, consisting of all measurable functions f on Ω such that for some $f_i \in X_i$ with $\|f_i\|_{X_i}$, $i = 0, 1$, and some $\lambda > 0$, it holds that

$$|f| \leq \lambda\psi(|f_0|, |f_1|) \quad \mu\text{–a.e. on } \Omega. \tag{8.14}$$

The norm of an element $f \in \varphi(\bar{X})$ is defined as

$$\|f\|_{\varphi} = \inf \lambda,$$

where the infimum is taken over all $\lambda > 0$ such that (8.14) holds. The space $\varphi(\bar{X})$ is a Banach lattice intermediate between X_0 and X_1; more precisely,

$$\|f\|_{\varphi} \leq \frac{1}{\varphi(1)} \|f\|_{\Delta(\bar{X})}, \quad f \in \Delta(\bar{X})$$

and

$$\|f\|_{\Sigma(\bar{X})} \leq \varphi(1) \|f\|_{\varphi}, \quad f \in \Sigma(\bar{X})$$

(cf. Maligranda [61], pp. 176–178).

In the case $\varphi(t) = t^{\theta}$, we denote $\varphi(\bar{X})$ by $X_0^{1-\theta} X_1^{\theta}$.

Theorem 8.6 If Φ_0 and Φ_1 are Orlicz functions, and

$$\Phi^{-1} = \psi(\Phi_0^{-1}, \Phi_1^{-1})$$

for some $\varphi \in \mathcal{P}$, then $\varphi(L^{\Phi_0}, L^{\Phi_1}) = L^{\Phi}$, and

$$\|f\|_{\varphi} \leq \|f\|_{\Phi} \leq 2 \|f\|_{\varphi}.$$

Moreover, $(L_{p_0})^{1-\theta}(L_{p_1})^{\theta} = L_p$ with equality of norms, where

$$\frac{1}{p} = \frac{1-\theta}{p_0} + \frac{\theta}{p_1}.$$

For a proof of this theorem, see e.g. Maligranda [61], pp. 179–180 or Ovchinnikov [71], pp. 460–461.

The Calderón–Lozanovskiĭ construction $\varphi(\cdot)$ is not, in general, an interpolation method for Banach lattices (see Lozanovskiĭ [59]). In 1976, V. I. Ovchinnikov [70] proved that if $\varphi \in \mathcal{P}$, then $\varphi(\cdot)$ is an interpolation method on couples of Banach lattices with the Fatou property (the interpolation constant was not specified). Then Gustavsson–Peetre [31], E. I. Berezhnoĭ [10] and V. A. Shestakov [78] proved that if $\varphi \in \mathcal{P}_{\pm}$, then $\varphi(\cdot)^c$ and $\varphi(\cdot)^0$ are interpolation methods on couples of Banach lattices[1]. Their interpolation constant depends on φ and increases to ∞ when we go with φ to the "boundaries" (cf. Remark 8.2); this follows from the fact that they used the Carlson type inequality (8.9).

To compare the Calderón–Lozanovskiĭ construction, which is not an interpolation method in general, with some other interpolation construction, we need the notion of orbits, introduced by N. Aronszajn and E. Gagliardo in 1965 [3]. Let $\bar{E} = (E_0, E_1)$ be a fixed Banach couple, and let E be an intermediate space between E_0 and E_1. For any Banach couple $\bar{A} = (A_0, A_1)$, the *orbit* of E in $\Sigma(\bar{A})$ is the space of all $a \in \Sigma(\bar{A})$ which admit a representation

$$a = \sum_{k=1}^{\infty} T_k(e_k) \quad \text{with convergence in } \Sigma(\bar{A}),$$

where $T_k \in L(\bar{E}, \bar{A})$ and

$$\sum_{k=1}^{\infty} \|T_k\|_{\bar{E}\to\bar{A}} \|e_k\|_E < \infty.$$

This space, denoted by $\mathrm{Orb} = \mathrm{Orb}_E^{\bar{E}}(\bar{A})$, is a Banach space with the norm

$$\|a\|_{\mathrm{Orb}} = \inf \sum_{k=1}^{\infty} \|T_k\|_{\bar{E}\to\bar{A}} \|e_k\|_E ,$$

where the infimum is taken over all admissible representations. It is easy to see that $\mathrm{Orb}_E^{\bar{E}}(\bar{A})$ is an exact interpolation method, which means that if $T : \bar{A} \to \bar{B}$, then

$$T : \mathrm{Orb}_E^{\bar{E}}(\bar{A}) \to \mathrm{Orb}_E^{\bar{E}}(\bar{B})$$

and

$$\|T\|_{\mathrm{Orb}\to\mathrm{Orb}} \le \|T\|_{\bar{A}\to\bar{B}} .$$

[1]For a Banach space A intermediate between A_0 and A_1, we denote by A^0 the closure of $\Delta(\bar{A})$ in A, and by A^c the *Gagliardo completion* of A with respect to $\Sigma(\bar{A})$, i.e. $a \in A^c$ if and only if there exists a sequence $\{a_k\}$ of elements in A such that, for some $\lambda < \infty$, $\|a_k\|_A \le \lambda$ and $\lim_{k\to\infty} \|a_k - a\|_{\Sigma(\bar{A})} = 0$. A^c is a Banach space with the norm $\|a\|_{A^c} = \inf \lambda$, where the infimum is taken over all admissible λ.

Moreover, if an exact interpolation method G is such that

$$E \overset{1}{\subseteq} G(\bar{E}),$$

then

$$\mathrm{Orb}_E^{\bar{E}}(\bar{A}) \overset{1}{\subseteq} G(\bar{A})$$

for every Banach couple $\bar{A}$. Our main interest in this construction is when

$$\bar{E} = \bar{c}_0 = (c_0, c_0(2^{-k})), \quad E = c_0(\{1/\varphi(2^k)\}),$$

where $\varphi \in \mathcal{P}_0$ and

$$\left\|\{a_k\}_{k=-\infty}^{\infty}\right\|_{c_0} = \max_{k \in \mathbb{Z}} |a_k| \quad \text{and} \quad \left\|\{a_k\}_{k=-\infty}^{\infty}\right\|_{c_0(2^{-k})} = \left\|\{a_k 2^{-k}\}_{k=-\infty}^{\infty}\right\|_{c_0}.$$

This orbit construction was introduced in 1971 by Peetre in another equivalent form and we denote it by $G_{\varphi,0}$. Thus Peetre's interpolation method is defined by the formula

$$G_{\varphi,0} = \mathrm{Orb}_{c_0}^{\bar{c}_0}(\{1/\varphi(2^k)\}).$$

If we replace c_0 by l_∞ everywhere, then we obtain the corresponding method

$$G_{\varphi,\infty} = \mathrm{Orb}_{l_\infty}^{\bar{l}_\infty}(\{1/\varphi(2^k)\}).$$

V. I. Ovchinnikov [71] proved that if $\varphi \in \mathcal{P}_0$, then

$$G_{\varphi,0}(\bar{X}) \subseteq \varphi(\bar{X}) \subseteq G_{\varphi,\infty}(\bar{X}) \subseteq \varphi(\bar{X})^c,$$

and the fact that $\varphi(\cdot)^c$ is an interpolation method for Banach lattices then follows directly (and the interpolation constant is independent of φ). In 1985, P. Nilsson [68] showed that if $\varphi \in \mathcal{P}_0 \cap \mathcal{P}_\infty$, then $G_{\varphi,0}(\bar{X}) = \varphi(\bar{X})^0$ and

$$\varphi(\bar{X}) \subseteq G_{\varphi,\infty}(\bar{X}) \subseteq \varphi(\bar{X})^c$$

even for quasi-Banach lattices. This gives the results that if $\varphi \in \mathcal{P}_0 \cap \mathcal{P}_\infty$, then $\varphi(\cdot)^c$ and $\varphi(\cdot)^0$ are interpolation methods for Banach lattices (and the interpolation constants are independent of φ). Ovchinnikov and Nilsson used some generalization of the Carlson inequality, but the proofs of these generalizations are not so clear. They also use some facts from interpolation theory, such as K-monotonicity from $\bar{l}_2$ to $\bar{l}_\infty$ and the K-divisibility property. Kruglyak–Maligranda–Persson [47] gave a clear proof for $\varphi \in \mathcal{P}_0$ by using the Carlson type inequality with blocks. This gives the interpolation result with a universal constant in the estimation of the operator norms (see also Brudnyĭ–Kruglyak [18], pp. 560–564). The main interpolation result in the Kruglyak–Maligranda–Persson paper [47] (see also previous papers by V. I. Ovchinnikov [71] and P. Nilsson [68]) was the following identification of the Calderón–Lozanovskiĭ construction with the Peetre method.

Theorem 8.7 If $\varphi \in \mathcal{P}_0$, then, for any couple of Banach lattices $\bar{X}$,

$$G_{\varphi,0}(\bar{X}) = \varphi(\bar{X})^0$$

and

$$G_{\varphi,0}(\bar{X})^c = \varphi(\bar{X})^c.$$

The embedding constants in both identifications are independent of φ and $\bar{X}$.

Remark 8.8 For the special case when $\varphi \in \mathcal{P}_0 \cap \mathcal{P}_\infty$ and $\bar{X}$ is a regular couple of lattices, this theorem was proved by Nilsson [68].

The proof of Theorem 8.7 will follow from the following lemmas, which are of independent interest. In the proof of Lemma 8.3, we use our block version of Carlson's inequality in a crucial way.

Lemma 8.3 If $\varphi \in \mathcal{P}_0$, then, for any couple $\bar{X}$ of Banach lattices,

$$G_{\varphi,0}(\bar{X}) \subseteq \varphi(\bar{X}),$$

where the embedding constant does not exceed $(1+\sqrt{2})^2$.

Proof. Step 1. It is sufficient to prove that for any interval $(a,b) \subset \mathbb{R}_+$ and $T : \bar{c}_0 \to \bar{X}$ we have, for

$$\varphi\chi_{(a,b)} = \sum \varphi\chi_{(a,b)}(2^k)e_k,$$

that

$$\left\|T(\varphi\chi_{(a,b)})\right\|_{\varphi(\bar{X})} \leq (1+\sqrt{2})^2 \, \|T\|_{\bar{c}_0 \to \bar{X}} \, . \tag{8.15}$$

In fact, since $\varphi(\bar{c}_0) = c_0(\{1/\varphi(2^k)\})$, it follows from (8.15) that, for any $f \in \varphi(\bar{c}_0)$ with finite support, we have

$$\|T(f)\|_{\varphi(\bar{X})} \leq (1+\sqrt{2})^2 \, \|T\|_{\bar{c}_0 \to \bar{X}} \, \|f\|_{\varphi(\bar{c}_0)} \, . \tag{8.16}$$

If $f \in \varphi(\bar{c}_0)$ has infinite support, then f can be written as a sum

$$f = \sum_{k=1}^{\infty} f_k,$$

of functions with finite support and such that

$$\sum_{k=1}^{\infty} \|f_k\|_{\varphi(\bar{c}_0)} \leq (1+\epsilon)\|f\|_{\varphi(\bar{c}_0)}, \quad \epsilon > 0.$$

The series $\sum Tf_k$ converges absolutely to Tf in $\Sigma(\bar{X})$ and, according to (8.16) it also converges absolutely in $\varphi(\bar{X})$, so we conclude that it converges absolutely to Tf in $\varphi(\bar{X})$. Moreover,

$$\begin{aligned}
\|Tf\|_{\varphi(\bar{X})} &\leq (1+\sqrt{2})^2 \|T\|_{\bar{c}_0 \to \bar{X}} \left(\sum_{k=1}^{\infty} \|f_k\|_{\varphi(\bar{c}_0)}\right) \\
&\leq (1+\epsilon)(1+\sqrt{2})^2 \|T\|_{\bar{c}_0 \to \bar{X}} \|f\|_{\varphi(\bar{c}_0)}.
\end{aligned}$$

Let $\epsilon \to 0^+$ and we find that (8.16) holds for *all* $f \in \varphi(\bar{c}_0)$. Now, if $f \in G_{\varphi,0}(\bar{X})$, then $f = \sum_{k=1}^{\infty} T_k f_k$ and

$$\begin{aligned}
\|f\|_{\varphi(\bar{X})} &= \left\|\sum_{k=1}^{\infty} T_k f_k\right\|_{\varphi(\bar{X})} \\
&\leq \sum_{k=1}^{\infty} \|T_k f_k\|_{\varphi(\bar{X})} \\
&\leq (1+\sqrt{2})^2 \sum_{k=1}^{\infty} \|T_k\|_{\bar{c}_0 \to \bar{X}} \|f_k\|_{\varphi(\bar{c}_0)},
\end{aligned}$$

i.e.

$$\|f\|_{\varphi(\bar{X})} \leq (1+\sqrt{2})^2 \|f\|_{G_{varphi,0}}.$$

Second Step. Let $\varphi = \varphi\chi_{(a,b)}$, $0 < a < b < \infty$, and consider, for $\chi_k = [t_{2k}, t_{2k+k})$

$$g_0 = \max_k \frac{|T(\varphi\chi_k)|}{\varphi(t_{2k+1})} \quad \text{and} \quad g_1 = \max_k \frac{t_{2k+1}}{\varphi(t_{2k+1})}|T(\varphi\chi_k)|.$$

We will prove that

$$\|g_i\|_{X_i} \leq r\|T\|_{\bar{c}_0 \to \bar{X}}, \quad i = 0, 1. \tag{8.17}$$

Since φ has finite support and $t_k \to 0$ as $k \to -\infty$ and $t_k \to \infty$ as $k \to \infty$, it follows that the maximum in both expressions defining g_0 and g_1 can be taken over only a finite number of indices, say m. According to the well-known inequality

$$\max_{1\le k\le m} |x_k| \le 2^{-m} \sum_{\epsilon_k=\pm 1} \left| \sum_{k=1}^{m} \epsilon_k x_k \right|$$

we have that

$$\begin{aligned} g_0 &\le 2^{-m} \sum_{\epsilon_k=\pm 1} \left| \sum_{k=1}^{m} \frac{\epsilon_k}{\varphi(t_{2k+1})} T(\varphi\chi_k) \right| \\ &= 2^{-m} \sum_{\epsilon_k=\pm 1} \left| T\left(\sum_{k=1}^{m} \frac{\epsilon_k}{\varphi(t_{2k+1})} \varphi\chi_k \right) \right| \end{aligned}$$

and thus

$$\|g_0\|_{X_0} \le 2^{-m} \sum_{\epsilon_k=\pm 1} \|T\|_{\bar{c}_0 \to \bar{X}} \left\| \sum_{k=1}^{\infty} \frac{\epsilon_k}{\varphi(t_{2k+1})} \varphi\chi_k \right\|_{\bar{c}_0} .$$

Since $\varphi\chi_k$ have disjoint supports for different k and $|\varphi\chi_k| \le r\varphi(t_{2k+1})$, it follows that

$$\|g_0\|_{X_0} \le r \, \|T\|_{\bar{c}_0 \to \bar{X}} \cdot$$

Similarly, we find that

$$\|g_1\|_{X_1} \le r \, \|T\|_{\bar{c}_0 \to \bar{X}} ,$$

and (8.17) is proved.

Third Step. According to our generalized Carlson inequality (8.13), we have

$$\begin{aligned} |T\varphi| &\le \sum_k |T\varphi\chi_k| \\ &= \sum_k \psi\left(\frac{|T\varphi\chi_k|}{\varphi(t_{2k+1})}, \frac{t_{2k+1}}{\varphi(t_{2k+1})} |T\varphi\chi_k| \right) \\ &\le \frac{r+1}{r-1} \psi(g_0, g_1). \end{aligned}$$

Therefore, $T\varphi \in \varphi(\bar{X})$ and, in view of (8.17), we conclude that

$$\|T\varphi\|_{\varphi(\bar{X})} \le \frac{r(r+1)}{r-1} \|T\|_{\bar{c}_0 \to \bar{X}} \cdot$$

By taking infimum over $r > 1$, we obtain the estimate (8.15), and the proof is complete. $\square$

Remark 8.9 In the first step of the proof, we haven't used the structure of $\bar{X}$ and $\varphi(\bar{X})$.

Lemma 8.4 If $\varphi \in \mathcal{P}_0$, then, for any couple of Banach lattices $\bar{X}$,

$$\varphi(\bar{X}) \subseteq G_{\varphi,0}(\bar{X}),$$

where the embedding constant does not exceed 2.

Proof. Let $\|f\|_{\varphi(\bar{X})} \leq 1$, i.e. there exist $f_i \in X_i$, such that $|f| \leq \psi(|f_0|, |f_1|)$ and $\|f_i\|_{X_i} \leq 1$, $i = 0, 1$. Let

$$\Omega_k = \left\{\omega \in \Omega; 2^k \leq \frac{|f_1(\omega)|}{|f_0(\omega)|} < 2^{k+1}\right\}, \quad k \in \mathbb{Z}.$$

Consider the mapping $A : \bar{l}_\infty \to \bar{X}$ such that $Ae_k = f\chi_{\Omega_k}$ and make an extension to $\Sigma(\bar{l}_\infty)$ formally defined by

$$A\left(\sum_k \lambda_k e_l\right) = \sum_k \lambda_k Ae_k.$$

Then $A\varphi = f$ (in this step it is important that $\varphi \in \mathcal{P}_0$) and $\|A\|_{\bar{l}_\infty \to \bar{X}} \leq 2$. $\square$

Lemma 8.5 If $\varphi \in \mathcal{P}_0$, then, for any Banach couple $\bar{X}$,

(*i*) $G_{\varphi,\infty}(\bar{X})^0 \overset{5}{\subseteq} G_{\varphi,0}(\bar{X}) \overset{1}{\subseteq} G_{\varphi,\infty}(\bar{X})^0$ and
(*ii*) $G_{\varphi,\infty}(\bar{X}) \overset{1}{\subseteq} G_{\varphi,0}(\bar{X})^c$.

Remark 8.10 For $\varphi \in \mathcal{P}_0 \cap \mathcal{P}_1$, (i) was proved by S. Janson [37].

Proof of Lemma 8.5. (i) The second embedding follows in the following way. If $\|f\|_{\varphi(\bar{c}_0)} \leq 1$, then the operator $A : \bar{l}_\infty \to \bar{c}_0$ defined by $A_f(e_k) = f(2^k)/\varphi(2^k)$ has norm

$$\|A\|_{\bar{l}_\infty \to \bar{c}_0} \leq 1.$$

Therefore, if $\|x\|_{G_{\varphi,0}} < 1$, then x has a representation $x = \sum_{k=1}^{\infty} T_k f_k$, with

$$\sum_{k=1}^{\infty} \|T_k\|_{\bar{c}_0 \to \bar{X}} \|f_k\|_{\varphi(\bar{c}_0)} < 1,$$

and thus $x = \sum_{k=1}^{\infty} T_k A_{f_k} \varphi$ with

$$\sum_{k=1}^{\infty} \|T_k A_{f_k}\|_{\bar{l}_\infty \to \bar{X}} \|\varphi\|_{l_\infty(\{1/\varphi(2^k)\})} < 1,$$

and the second embedding is proved. In order to prove the first embedding, it is sufficient to prove that

$$\|x\|_{G_{\varphi,0}(\bar{X})} \leq 5 \|x\|_{G_{\varphi,\infty}(\bar{X})}, \quad x \in \Delta(\bar{X}). \tag{8.18}$$

Let $x \in \Delta(\bar{X})$ and $\|x\|_{G_{\varphi,0}(\bar{X})} < 1$. Then $x = A\varphi$ for some $A : \bar{l}_\infty \to \bar{X}$ with $\|A\|_{\bar{l}_\infty \to \bar{X}} < 1$. We have

$$x = A(\varphi \chi_{(-\infty,-m)}) + A(\varphi \chi_{[-m,m]}) + A(\varphi \chi_{(m,\infty)}).$$

First, we note that $\|\varphi \chi_{[-m,m]}\|_{\varphi(\bar{c}_0)} \leq 1$ and since the restriction of A to $\bar{c}_0$ is an operator with norm < 1 it follows that

$$\|A(\varphi \chi_{[-m,m]})\|_{G_{\varphi,0}} < 1.$$

Therefore it is sufficient to prove that for an arbitrary $\epsilon > 0$ and for sufficiently large m, we have the estimates

$$\|A(\varphi \chi_{(-\infty,-m)})\|_{G_{\varphi,0}} \leq 2 + \epsilon$$

and

$$\|A(\varphi \chi_{(m,\infty)})\|_{G_{\varphi,0}} \leq 2 + \epsilon.$$

The proofs of these estimates are quite similar, so we only prove, for example, the second one. Let us consider an operator $B : \bar{c}_0 \to \bar{X}$ such that $B(e_k) = 0$ if $k \neq m$ and

$$B(\varphi(2^m) e_m) = A(\varphi \chi_{(m,\infty)}).$$

We need to prove that for large m, we have

$$\|B\|_{\bar{c}_0 \to \bar{X}} \leq 2 + \epsilon,$$

i.e.

$$\|A(\varphi \chi_{(m,\infty)})\|_{X_0} \leq (2 + \epsilon)\varphi(2^m) \tag{8.19}$$

and

$$\|A(\varphi \chi_{(m,\infty)})\|_{X_1} \leq (2 + \epsilon) 2^{-m} \varphi(2^m). \tag{8.20}$$

Since φ is concave and $\|A\|_{\bar{l}_\infty \to \bar{X}} < 1$, it follows that

$$\left\|A(\varphi\chi_{(m,\infty)})\right\|_{X_1} \le \left\|\varphi\chi_{(m,\infty)}\right\|_{l_\infty(2^{-k})} \le 2^{-m}\varphi(2^m),$$

and the inequality (8.19) holds (even with constant 1). For the proof of (8.20), we consider the following cases:

1^0 $\lim_{m\to\infty}\varphi(2^m) = \infty$. We choose m so large that $\|A\varphi\|_{X_0} < \varphi(2^m)$ and find that

$$\left\|A(\varphi\chi_{(m,\infty)})\right\|_{X_0} \le \|A\varphi\|_{X_0} + \left\|A(\varphi\chi_{(-\infty,m)})\right\|_{X_0} < 2\varphi(2^m).$$

2^0 $\lim_{m\to\infty}\varphi(2^m) = C < \infty$. In this case, we have that $\|\varphi\|_{l_\infty} = C$ and $\left\|\varphi\chi_{(-\infty,m)}\right\|_{l_\infty} \le C$. Therefore

$$\left\|A(\varphi\chi_{(m,\infty)})\right\|_{X_0} \le \|A\varphi\|_{X_0} + \left\|A(\varphi\chi_{(-\infty,m)})\right\|_{X_0} < 2C$$

and for large m we also have $2C < (2+\epsilon)\varphi(2^m)$. Thus (8.20) holds, which yields (8.18), so (i) is proved.

(ii) This follows from the minimality of $G_{\varphi,\infty}$ in the couple $\bar{l}_\infty$ and the fact that $G_{\varphi,0}(\bar{X})$ is an exact interpolation method, the Gagliardo completion on the couple $\bar{l}_\infty$ of which contains φ (since for the embedding operator $A : \bar{c}_0 \to \bar{l}_\infty$ we have $A(\varphi(\bar{c}_0)) = c_0(\{1/\varphi(2^k)\})$ with Gagliardo completion $l_\infty(\{1/\varphi(2^k)\})$). □

We are now ready to prove the main result of this section.

Proof of Theorem 8.7. (a) By using Lemmas 8.3–8.5, and $G_{\varphi,0}(\bar{X}) \equiv G_{\varphi,0}(\bar{X})^0$ (this equality with equal norms follows from Lemma 8.5 (i)), we find that

$$\varphi(\bar{X})^0 \overset{2}{\subseteq} G_{\varphi,\infty}(\bar{X})^0 \overset{5}{\subseteq} G_{\varphi,0}(\bar{X}) \equiv G_{\varphi,0}(\bar{X})^0 \overset{(1+\sqrt{2})^2}{\subseteq} \varphi(\bar{X})^0,$$

and we have in particular proved that

$$G_{\varphi,0}(\bar{X}) = \varphi(\bar{X})^0.$$

(b) According to Lemmas 8.3, 8.4 and 8.5 (ii), we have that

$$G_{\varphi,0}(\bar{X}) \overset{(1+\sqrt{2})^2}{\subseteq} \varphi(\bar{X}) \overset{2}{\subseteq} G_{\varphi,\infty}(\bar{X}) \overset{1}{\subseteq} G_{\varphi,0}(\bar{X})^c,$$

and this yields that $G_{\varphi,0}(\bar{X})^c = \varphi(\bar{X})^c$. □

We close this section by stating the following interpolation results.

Corollary 8.1 If $\varphi \in \mathcal{P}_0$, then $\varphi(\cdot)^0$ and $\varphi(\cdot)^c$ are interpolation methods on couples of Banach lattices with interpolation constants not exceeding $10(1+\sqrt{2})^2$ and $2(1+\sqrt{2})^2$, respectively.

We can also prove the following result on interpolation of Orlicz spaces (see Maligranda [61], Theorem 14.12).

Theorem 8.8 Let Φ_0, Φ_1 and Ψ_0, Ψ_1 be Orlicz functions. If

$$T : (L^{\Phi_0}(\mu), L^{\Phi_1}(\mu)) \to (L^{\Psi_0}(\nu), L^{\Psi_1}(\nu))$$

and

$$\Phi^{-1} = \Phi_0^{-1}\varphi\left(\frac{\Phi_1^{-1}}{\Phi_0^{-1}}\right), \quad \Psi^{-1} = \Psi_0^{-1}\varphi\left(\frac{\Psi_1^{-1}}{\Psi_0^{-1}}\right)$$

with $\varphi \in \mathcal{P}$, then $T : L^{\Phi}(\mu) \to L^{\Psi}(\nu)$ and

$$\|T\|_{L^\Phi \to L^\Psi} \leq 26 \max_{i=0,1}\{\|T\|_{L^{\Phi_i} \to L^{\Psi_i}}\}. \tag{8.21}$$

Moreover, if $\varphi \in \mathcal{P}_0$, then the inequality (8.21) holds with constant 12.

Remark 8.11 Careful analysis of the proofs show that, in fact, the constant 26 in (8.21) can be replaced by

$$C \leq 2(3+2\sqrt{2})C_\Psi < 12C\Psi,$$

where

$$C_\Psi = \sup_{t>0} \frac{\Psi^{-1}(2t)}{\Psi^{-1}(t)} \leq 2$$

(see Karlovich–Maligranda [41], p. 2728).

Remark 8.12 Theorem 8.8 for positive operators T holds with constant 1 in place of 26, which was proved independently by E. I. Berezhnoi [10], V. A. Shestakov [78] and Maligranda [60] (cf. [61], Theorem 15.13).

Chapter 9

Related Results and Applications

In this chapter, we present some related results, applications and open questions that do not seem to fit in elsewhere in the text, but which can be of importance for various reasons (e.g. as a source of inspiration for further research).

9.1 A Generalization of Redheffer

If u is a real-valued function on $\mathbb{R}^n$, we define a non-negative function of $r > 0$ by

$$[u]^2(r) = \frac{1}{\sigma_n r^{n-1}} \int_{|x|=r} |u(x)|^2 \, d\sigma,$$

where σ denotes the surface area measure on the sphere $|x| = r$ in $\mathbb{R}^n$ and σ_n is the area of the unit sphere in $\mathbb{R}^n$. R. Redheffer [75] proved a theorem, of which we state the following special case.

Theorem 9.1 Suppose that u is differentiable on some region in $\mathbb{R}^n$, and suppose that u and H are such that $[\nabla u]^2$ and $[u]^2/H$ are integrable on (a, b). Define

$$F(r) = \int_{r_0}^{r} [u]^2 \, dr - [u]^2\Big|_{r_0}^{r},$$

where $r_0 \in (a, b)$. Then F is of bounded variation on (a, b), and

$$|F(b) - F(a)| \le 2 \left(\int_a^b [\nabla u]^2 \, dr \right)^{\frac{1}{2}} \left(\int_a^b [u]^2 \frac{dr}{H} \right)^{\frac{1}{2}}. \tag{9.1}$$

If, moreover, $H > 0$, then there is equality only when $[u]/H \equiv 0$ or $u = ce^{d/H}$ for some constants c and d.

He also pointed out that Carlson's inequality (1.1) follows from a special case of this result. Indeed, letting $n = 1$, $a = 0$, $b = \frac{\pi}{2}$, and $H \equiv 1$, and taking

$$u(x) = a_1 \cos x + a_2 \cos 3x + \ldots + a_k \cos(2k-1)x$$

we find that

$$\begin{aligned}
[u](0) &= \sum_{k=1}^{m} a_k, \\
[u]\left(\frac{\pi}{2}\right) &= 0, \\
F(0) - F\left(\frac{\pi}{2}\right) &= \left(\sum_{k=1}^{m} a_k\right)^2, \\
[\nabla u]^2(r) &= \left(\sum_{k=1}^{m} (2k-1) a_k \sin(2k-1)r\right)^2, \\
\int_0^{\pi/2} [\nabla u]^2(r)\, dr &= \pi \sum_{k=1}^{m} \left(k - \frac{1}{2}\right)^2 a_k^2, \\
[u]^2(r) &= \left(\sum_{k=1}^{m} a_k \cos(2k-1)r\right)^2, \\
\int_0^{\pi/2} [u]^2(r)\, dr &= \frac{\pi}{4} \sum_{k=1}^{m} a_k^2,
\end{aligned}$$

and, thus, (9.1) implies

$$\begin{aligned}
\sum_{k=1}^{m} a_k &< \sqrt{2}\left(\pi \sum_{k=1}^{m} \left(k - \frac{1}{2}\right)^2 a_k^2\right)^{\frac{1}{4}} \left(\frac{\pi}{4} \sum_{k=1}^{m} a_k^2\right)^{\frac{1}{4}} \\
&= \sqrt{\pi}\left(\sum_{k=1}^{m} \left(k - \frac{1}{2}\right)^2 a_k^2\right)^{\frac{1}{4}} \left(\sum_{k=1}^{m} a_k^2\right)^{\frac{1}{4}}.
\end{aligned}$$

Since the bound is uniform in m, we may let $m \to \infty$, which gives Landau's inequality (see Section 2.6), and Carlson's inequality (1.1) certainly follows from this.

9.2 Sobolev Type Embeddings

Barza et al. [5] applied their Theorem 5.1 to Fourier transforms of functions on $\mathbb{R}^n$ in order to prove embeddings of Sobolev type (cf. also Beurling [15] and Sz. Nagy [82]). If $f : \mathbb{R}^n \to \mathbb{C}$ is integrable, we define the Fourier transform Ff of f by

$$Ff(\xi) = (2\pi)^{-n/2} \int_{\mathbb{R}^n} e^{-i\langle x,\xi\rangle} f(x)\, dx.$$

The inverse of $F : L_2 \to L_2$ is given by the integral formula

$$F^{-1}g(x) = (2\pi)^{-n/2} \int_{\mathbb{R}^n} e^{i\langle x,\xi\rangle} g(\xi)\, d\xi.$$

Also, for any positive integer l, define

$$\nabla^l f = \left(\frac{\partial^l f}{\partial x_{i_1} \cdots \partial x_{i_l}} \right)^n_{i_1,\ldots,i_l=1}.$$

Theorem 9.2 If $l > n/2$, then

$$\|f\|_{L_\infty(\mathbb{R}^n)} \le C \|f\|^{1-n/2l}_{L_2(\mathbb{R}^n)} \left\| |\nabla^l f| \right\|^{n/2l}_{L_2(\mathbb{R}^n)}$$

for any f such that the norms on the right-hand side are finite. We can choose $C > 0$ with

$$C^2 = \frac{2^{-n}\pi^{1-n/2}}{l\Gamma\left(\frac{n}{2}\right) \sin \frac{n\pi}{2l}} \left(1 - \frac{n}{2l}\right)^{-(1-\frac{n}{2l})} \left(\frac{n}{2l}\right)^{-\frac{n}{2l}}.$$

Proof. By Theorem 5.1 with

$$w(\xi) = |\xi|^n, \quad w_0(\xi) = |\xi|^{\frac{n}{2}}, \quad w_1(\xi) = |\xi|^{l+\frac{n}{2}},$$

$$p = 1, \quad p_0 = p_1 = 2,$$

$$\theta = \frac{n}{2l}$$

and the Parseval identity

$$\begin{aligned}
|f(x)| &= |F^{-1}Ff(x)| \\
&\leq (2\pi)^{-n/2} \int_{\mathbb{R}^n} |Ff(\xi)||\xi|^n \frac{d\xi}{|\xi|^n} \\
&\leq (2\pi)^{-n/2} C_0 \left(\int_{\mathbb{R}^n} |Ff(\xi)|^2 \, d\xi \right)^{\frac{1}{2}(1-\frac{n}{2l})} \left(\int_{\mathbb{R}^n} |\xi|^{2l} |Ff(\xi)|^2 \, d\xi \right)^{\frac{1}{2}\frac{n}{2l}} \\
&= (2\pi)^{-n/2} C_0 \, \|Ff\|_{L_2(\mathbb{R}^n)}^{1-\frac{n}{2l}} \, \||F(\nabla^l f)|\|_{L_2(\mathbb{R}^n)}^{\frac{n}{2l}} \\
&= (2\pi)^{-n/2} C_0 \|f\|_{L_2(\mathbb{R}^n)}^{1-\frac{n}{2l}} \, \||\nabla^l f|\|_{L_2(\mathbb{R}^n)}^{\frac{n}{2l}} ,
\end{aligned}$$

where C_0 is the constant in Theorem 5.1 with the parameters replaced by the values given here. □

9.3 A Local Hausdorff–Young Inequality

Kamaly [40] used his inequality from Theorem 5.8 to prove a sharp version of a Hausdorff–Young type inequality for functions with small supports. We would like to illustrate the wide range of applicability of inequalities of Carlson type by stating this result.

The Hausdorff–Young inequality is well-known: if $p \in [1, 2]$, then for all $f \in L_p(\mathbb{T})$ it holds that

$$\|\hat{f}\|_{l_{p'}(\mathbb{Z})} \leq C \, \|f\|_{L_p(\mathbb{T})} , \tag{9.2}$$

where $C = 1$ is the best constant. In the case of the real line, we have

$$\|\hat{f}\|_{L_{p'}(\mathbb{R})} \leq C \, \|f\|_{L_p(\mathbb{R})} ,$$

and the best constant is $C = B_p$, where B_p is the Babenko–Beckner constant, given by

$$B_p = \sqrt{\frac{p^{\frac{1}{p}}}{p'^{\frac{1}{p'}}}}$$

(see [8]). Now, consider the n-dimensional torus $\mathbb{T}^n$, and, for $a > 0$, let

$$H_{p,r} = \sup \left\{ \frac{\|\hat{f}\|_{p'}}{\|f\|_p} ; f \in L_p(\mathbb{T}^n), \operatorname{supp} f \subseteq \bar{B}(0, r), f \neq 0 \right\}.$$

Thus, $H_{p,r}$ is the best constant in (9.2) in the n-dimensional case, when we restrict our attention to functions with support contained in the closed ball centered at 0 with radius r. We put

$$H_p = \lim_{r \searrow 0} H_{p,r}.$$

Theorem 9.3 (Kamaly 2000) For $n \in \mathbb{Z}_+$, it holds that

$$H_p \leq B_p^n.$$

Thus, when considering only functions with small supports, the constant can be chosen considerably smaller than 1.

Remark 9.1 The corresponding result for the case $n = 1$ was previously proved by M. E. Andersson [1] in the case where p' is an even integer, and by P. Sjölin [79] in the general case.

9.4 Optimal Sampling

J. Bergh [11] applied Beurling's version (3.6) of Carlson's integral inequality (3.1) to an optimal sampling problem, in which the issue is to minimize the *pulse energy*

$$\int \varphi^2(x)\, dx$$

given a maximal *error energy*

$$\int_{k-1/2}^{k+1/2} (((f(x) - f(k))\varphi(x-k))^2\, dx. \tag{9.3}$$

Here, f is a function in the Schwartz class $\mathcal{S}$ whose Fourier transform $\hat{f}$ has support in $[-\frac{1}{2}, \frac{1}{2}]$. Moreover, φ is thought of as an approximation of the δ function; more precisely, φ is even and positive, with support in $[-\frac{1}{2}, \frac{1}{2}]$, and

$$\int \varphi(x)\, dx = 1.$$

The error energy measures how accurately the function $f(x)$ can be reconstructed from sample values $f(k)$.

Let

$$S(x) = \frac{\sin \pi x}{\pi x},$$

and let τ_k be the translation operator ($\tau_k\varphi(x) = \varphi(x-k)$ etc.). The Sampling Theorem then says that if f is as above, then f can be reconstructed by means of the formula

$$f = S * \sum_k f(k)\tau_k\delta,$$

where δ is the Dirac measure, and the sum is taken in $\mathcal{S}'$. Now, it is impossible to construct a Dirac measure in terms of electric circuits. We therefore consider the difference

$$S * \left(f\sum_k \tau_k\varphi - \sum_k f(k)\tau_k\varphi\right),$$

representing the error when we reconstruct the function f from its sample values $\{f(k)\}$.

The problem now reduces to estimating the error energy of the function in parentheses on an arbitrary sample interval $[k-\frac{1}{2}, k+\frac{1}{2}]$, thus considering the integral in (9.3). It does not exceed

$$\sup_x |f'(x)|^2 \int x^2\varphi^2(x)\,dx.$$

Suppose now that the error energy is allowed to reach at most $\epsilon > 0$. Since $\int \varphi(x)\,dx = 1$, Beurling's inequality (3.6) then implies the estimate

$$\int \varphi^2(x)\,dx \geq \frac{1}{4\pi^2\epsilon}\sup_x |f'(x)|^2$$

for the pulse energy. φ can now be taken close to a maximizing function for the inequality (we may choose a truncation of a dilation of $1/(1+x^2)$, multiplied by a suitable constant), thereby obtaining close to minimal pulse energy.

9.5 More on Interpolation, the Peetre Parameter Theorem

Initially (see e.g. [73]), the spaces of means, or *espaces de moyennes*, were defined using the three parameters θ, p_0 and p_1, as follows (see Chapter 7 for notation). If $\bar{A} = (A_0, A_1)$ is a compatible couple of Banach spaces, then

$$\bar{A}_{\theta,p_0,p_1}$$

is used to denote the space of $a \in \Sigma(\bar{A})$ for which there is a measurable function $u : \mathbb{R}_+ \to \Delta(\bar{A})$ such that

$$a = \int_0^\infty u(t) \frac{dt}{t} \quad \text{in } \Sigma(\bar{A}),$$

$$t \mapsto t^{-\theta} u(t) \in L^*_{p_0}(A_0)$$

and

$$t \mapsto t^{1-\theta} u(t) \in L^*_{p_1}(A_1).$$

Here, $L^*_p(A)$ is the space of functions with values in the Banach space A such that

$$\|f\|_{L^*_p(A)} = \left(\int_0^\infty \|f(t)\|_A^p \frac{dt}{t} \right)^{1/p} < \infty,$$

with the usual convention in the case $p = \infty$ (the asterisk merely indicates that we use the measure $\frac{dt}{t}$ rather than dt). Peetre [72] used Levin's theorem as a crucial step in the proof of the following result (the Peetre Parameter Theorem), reducing the number of parameters by one.

Theorem 9.4 (Peetre, 1963) If $0 < \theta < 1$, $1 \leq p_0, p_1 \leq \infty$, then

$$\bar{A}_{\theta, p_0, p_1} = \bar{A}_{\theta, p, p},$$

where

$$\frac{1}{p} = \frac{1-\theta}{p_0} + \frac{\theta}{p_1}.$$

Thus $\bar{A}_{\theta, p_0, p_1}$ may be a pre-version of the J-method in real interpolation theory.

If α is any real number, we denote by $L^*_{p,\alpha}$ the space determined by the finiteness of the norm

$$\|f\|_{L^*_{p,\alpha}} = \left(\int_0^\infty |s^{-\alpha} f(s)|^p \frac{ds}{s} \right)^{1/p}.$$

We will only prove one half of the following crucial lemma, which is a special case of Theorem 9.4, where Levin's theorem (Theorem 4.1) was successfully used in the proof.

Lemma 9.1 If

$$\frac{1}{p} = \frac{1-\theta}{p_0} + \frac{\theta}{p_1}$$

and

$$\alpha = (1-\theta)\alpha_0 + \theta\alpha_1,$$

then

$$(L^*_{p_0,\alpha_0}, L^*_{p_1,\alpha_1})_{\theta,p_0,p_1} = L^*_{p,\alpha}. \tag{9.4}$$

Proof of the inclusion $\subseteq$. Suppose that a is an element of the space on the left-hand side of (9.4), so that for some function u we have

$$a(s) = \int_0^\infty u(s,t)\frac{dt}{t},$$

$$(s,t) \mapsto t^{-\theta}s^{-\alpha_0}u(s,t) \in L^*_{p_0}(L^*_{p_0})$$

and

$$(s,t) \mapsto t^{1-\theta}s^{-\alpha_1}u(s,t) \in L^*_{p_1}(L^*_{p_1}).$$

According to Theorem 4.1, with $f(t)$ replaced by $u(s,t)/t$ and with parameters

$$\lambda = \theta p_0, \quad \mu = (1-\theta)p_1,$$

$$p = p_0, \quad q = p_1,$$

we get, by also using the Hölder–Rogers inequality,

$$\begin{aligned} |a(s)| &\le \int_0^\infty |u(s,t)|\frac{dt}{t} \\ &\le C\left(\int_0^\infty |t^{-\theta}u(s,t)|^{p_0}\frac{dt}{t}\right)^{\frac{1-\theta}{p_0}}\left(\int_0^\infty |t^{1-\theta}u(s,t)|^{p_1}\frac{dt}{t}\right)^{\frac{\theta}{p_1}}, \end{aligned}$$

and hence, in view of the definition of α

$$\begin{aligned} |s^{-\alpha}a(s)|^p \le C^p &\left(\int_0^\infty |t^{-\theta}s^{-\alpha_0}u(s,t)|^{p_0}\frac{dt}{t}\right)^{\frac{(1-\theta)p}{p_0}} \\ &\left(\int_0^\infty |t^{1-\theta}s^{-\alpha_1}u(s,t)|^{p_0}\frac{dt}{t}\right)^{\frac{\theta p}{p_1}}. \end{aligned}$$

By applying the Hölder–Rogers inequality once more, it thus follows that

$$\int_0^\infty |s^{-\alpha}a(s)|^p \frac{ds}{s} \le C^p \left(\int_0^\infty \int_0^\infty |t^{-\theta}s^{-\alpha_0}u(s,t)|^{p_0} \frac{ds}{s}\frac{dt}{t}\right)^{\frac{(1-\theta)p}{p_0}}$$
$$\left(\int_0^\infty \int_0^\infty |t^{1-\theta}s^{-\alpha_1}u(s,t)|^{p_1} \frac{ds}{s}\frac{dt}{t}\right)^{\frac{\theta p}{p_1}} < \infty.$$

We conclude that $s \mapsto s^{-\alpha}a(s) \in L_p^*$, or $a \in L_{p,\alpha}^*$. □

9.6 Carlson Type Inequalities with Several Factors

We give a brief description of a possible way to get Carlson type inequalities involving integrals on general measure spaces, as in Chapter 6, but with any finite number of factors on the right-hand side. The next theorem is presented here without proof (for more details, see also [48]). It should be mentioned that similar ideas also apply to more specific settings, such as on $\mathbb{R}_+$ or infinite cones in $\mathbb{R}^n$. Many of the classical results with more than two factors on the right-hand side follow from this, apart from the fact that it is not possible to give a sharp estimate of the constant in the general setting.

Theorem 9.5 Let $m \ge 2$ be an integer. Let (Ω, μ) be a measure space, on which the weights w, w_i, $i = 1, \ldots, m$, are defined. Suppose, moreover, that the parameters $p, p_i \in (0, \infty]$ and $\theta_i \in (0,1)$, $i = 1, \ldots, m$, are such that

$$\sum_{i=1}^m \theta_i = 1$$

and

$$\frac{1}{q} := \frac{1}{p} - \sum_{i=1}^m \frac{\theta_i}{p_i} \ge 0.$$

If, for some pair of indices i, j, with $1 \le i < j \le m$, it holds that

$$\mu\left(\left\{2^k \le \frac{w_j}{w_i} < 2^{k+1}\right\}\right) \le B, \quad k \in \mathbb{Z}$$

for some constant B, and if for some $s \in [q, \infty]$ we have

$$\frac{w}{w_1^{\theta_1} \cdots w_m^{\theta_m}} \in L_s(\Omega, \mu),$$

then there is a constant C such that

$$\|fw\|_{L_p(\Omega,\mu)} \leq C \prod_{i=1}^{m} \|fw_i\|_{L_{p_i}(\Omega,\mu)}^{\theta_i}.$$

9.7 Reverse Carlson Type Inequalities

We consider here the question of *reverse* inequalities of Carlson type, i.e. inequalities of the form

$$\|fw_0\|_{L_{p_0}}^{1-\theta} \|fw_1\|_{L_{p_1}}^{\theta} \leq C \|fw\|_{L_p}. \tag{9.5}$$

One may ask whether it is possible to find conditions on the weights such that the inequality (9.5) holds for some constant C, possibly under some additional conditions on the measure space (e.g. finiteness) or on the class of functions (such as monotonicity or convexity). We begin with a result which gives an affirmative answer to this question.

Theorem 9.6 Let (Ω, μ) be a measure space with $\mu(\Omega) < \infty$, and suppose that the parameters $0 < p_0, p_1 \leq p \leq \infty$ and $0 < \theta < 1$ are given. Suppose, moreover, that the weights w, w_0 and w_1 defined on Ω satisfy

$$\frac{w_i}{w} \in L_{q_i}(\Omega, \mu), \quad i = 0, 1, \tag{9.6}$$

where the q_i are defined by

$$\frac{1}{q_i} = \frac{1}{p_i} - \frac{1}{p}.$$

Then there is a constant C such that (9.5) holds.

Remark 9.2 Since the measure space is finite, the conditions (9.6) are implied by

$$\frac{w_i}{w} \in L_{s_i}(\Omega, \mu), \quad \text{some } s_i \in [q_i, \infty].$$

In particular, this is the case if the quotients w_i/w are essentially bounded.

Proof of Theorem 9.6. For $i = 0, 1$, we have by the Hölder–Rogers inequality

$$\begin{aligned}\|fw_i\|_{L_{p_i}}^{p_i} &= \int_\Omega |fw_i|^{p_i}\,d\mu \\ &= \int_\Omega |fw|^{p_i}\left(\frac{w_i}{w}\right)^{p_i} d\mu \\ &\le \left(\int_\Omega |fw|^p\,d\mu\right)^{\frac{p_i}{p}} \left(\int_\Omega \left(\frac{w_i}{w}\right)^{q_i} d\mu\right)^{\frac{p_i}{q_i}} \\ &= \left\|\frac{w_i}{w}\right\|_{L_{q_i}}^{p_i} \|fw\|_{L_{p_i}}^{p_i}.\end{aligned}$$

We see that (9.5) holds with

$$C = \left\|\frac{w_0}{w}\right\|_{L_{q_0}}^{1-\theta} \left\|\frac{w_1}{w}\right\|_{L_{q_1}}^{\theta}.$$

□

If we want to make an attempt to prove a reverse inequality on an infinite measure space, it is suggested by Levin's Theorem (Theorem 4.1) to consider an inequality of the form

$$\left(\int_0^\infty x^{p-1-\lambda} f^p(x)\,dx\right)^s \left(\int_0^\infty x^{q-1+\mu} f^q(x)\,dx\right)^t \le C\int_0^\infty f(x)\,dx, \quad (9.7)$$

where

$$s = \frac{\mu}{p\mu + q\lambda} \quad \text{and} \quad t = \frac{\lambda}{p\mu + q\lambda}.$$

It is natural to require $0 < p, q < 1$. However, the inequality (9.7) fails, even if we restrict our attention to the class of decreasing functions on $\mathbb{R}_+$.

Proposition 9.1 If $0 < p, q < 1$ and $\lambda, \mu > 0$, then there is a sequence $\{f_k\}_{k=1}^\infty$ of decreasing functions on $\mathbb{R}_+$ such that the sequence $\{Q_k\}_{k=1}^\infty$, defined by

$$Q_k = \frac{(\int_0^\infty x^{p-1-\lambda} f_k^p(x)\,dx)^s (\int_0^\infty x^{q-1+\mu} f_k^q(x)\,dx)^t}{\int_0^\infty f_k(x)\,dx}$$

tends to ∞ as $k \to \infty$.

Proof. Suppose first that $\lambda \ge p$. If f_0 is the characteristic function of the unit interval, then the integral on the right-hand side of (9.7), with $f = f_0$, equals 1, while

$$\int_0^\infty x^{p-1-\lambda} f_0^p(x)\,dx \geq \int_0^1 x^{-1}\,dx = \infty.$$

Thus, it suffices to put $f_k = f_0$ for all k. Suppose now that $0 < \lambda < p$. Let

$$f_k(x) = \begin{cases} k, & 0 < x < \frac{1}{k}, \\ x^{-1}, & \frac{1}{k} < x < k, \\ 0, & k < x < \infty, \end{cases} \qquad k = 1, 2, \ldots.$$

Then

$$\int_0^\infty f_k(x)\,dx = 1 + 2\log k,$$

$$\int_0^\infty x^{p-1-\lambda} f_k^p(x)\,dx = \frac{pk^\lambda}{\lambda(p-\lambda)} - \frac{k^{-\lambda}}{\lambda},$$

$$\int_0^\infty x^{q-1+\mu} f_k^q(x)\,dx = \frac{k^\mu}{\mu} - \frac{qk^{-\mu}}{\mu(q+\mu)}.$$

It follows that

$$Q_k = \frac{k^{\lambda s+\mu t}\left(\frac{p}{\lambda(p-\lambda)} - \frac{k^{-2\lambda}}{\lambda}\right)^s \left(\frac{1}{\mu} - \frac{qk^{-2\mu}}{\mu(q+\mu)}\right)^t}{1 + 2\log k}$$

$$= B_k \frac{k^{\frac{2\lambda\mu}{p\mu+q\lambda}}}{1 + 2\log k},$$

where B_k is bounded and bounded away from 0. Thus $Q_k \to \infty$ as $k \to \infty$. □

For related results, see also N. Leblanc [54].

9.8 Some Further Possibilities

In this book, we have pointed out a number of feasible further generalizations and also some open questions. We will here mention some further possibilities of this type.

9.8.1 *Other Function Spaces*

As has been mentioned earlier, inequalities of Carlson type can be thought of as inequalities between norms of certain normed spaces, which are in one way or another compatible:

$$\|f\|_X \leq C \|f\|_{A_0}^{1-\theta} \|f\|_{A_1}^{\theta}, \tag{9.8}$$

where $0 < \theta < 1$, or, more generally

$$\|f\|_X \leq C \prod_{j=1}^{m} \|f\|_{A_j}^{\theta_j}, \tag{9.9}$$

where $\sum_{j=1}^{m} \theta_j = 1$, respectively. It is then natural to initially ask: Does there exist other function spaces than weighted Lebesgue spaces (e.g. Orlicz spaces, Lorentz spaces, Besov spaces, etc.), where inequalities of the types (9.8) and (9.9) hold? If so, continue to develop the theory e.g. guided by the results and ideas presented in this book!

9.8.2 *Matrix Weights*

In a number of papers (see e.g. [83]), S. Treil and A. Volberg have considered Lebesgue type spaces with matrix weights. Is it possible to prove inequalities of Carlson type for this type of spaces? If so, is it then possible to obtain embeddings of real interpolation spaces, in the spirit of Chapter 7? S. Roudenko [77] has developed a theory for Besov spaces with matrix weights. Can we obtain Carlson type inequalities and embeddings of real interpolation spaces in this setting?

9.9 Necessity in the Case of a General Measure

It is of great interest to find some necessity results in the case of general measure spaces, although perhaps it is not possible in the most general case. Here, we only make the following concluding comments.

Let (Ω, μ) be a measure space, and let $p, p_0, p_1 \in (0, \infty]$ and $\theta \in (0, 1)$ satisfy suitable conditions. Also, suppose that the weights w, w_0, and w_1 are defined on Ω. We consider the inequality

$$\left(\int_\Omega |fw|^p \, d\mu \right)^{\frac{1}{p}} \leq C \left(\int_\Omega |fw_0|^{p_0} \, d\mu \right)^{\frac{1-\theta}{p_0}} \left(\int_\Omega |fw_1|^{p_1} \, d\mu \right)^{\frac{\theta}{p_1}}. \tag{9.10}$$

Now, make the change of measure $d\mu = \varphi \, d\nu$, and let

$$v = w\varphi^{\frac{1}{p}}, \quad v_i = w_i \varphi^{\frac{1}{p_i}}, i = 0, 1.$$

Then (9.10) is equivalent to

$$\left(\int_{\Omega}|fv|^{p}\,d\nu\right)^{\frac{1}{p}} \leq C\left(\int_{\Omega}|fv_0|^{p_0}\,d\nu\right)^{\frac{1-\theta}{p_0}}\left(\int_{\Omega}|fv_1|^{p_1}\,d\nu\right)^{\frac{\theta}{p_1}}. \qquad (9.11)$$

Now, the equivalent inequalities (9.10) and (9.11) have exactly the same shape. Therefore, any set of conditions on the triple (w, w_0, w_1) in terms of the measure μ and the parameters (p, p_0, p_1, θ) has to be equivalent to the corresponding set of conditions on the triple (v, v_0, v_1) in terms of ν and (p, p_0, p_1, θ), in order to be necessary and sufficient for the inequality (9.10) to hold. Thus, in this sense, any set of necessary and sufficient conditions for a Carlson type inequality must be *invariant under any absolutely continuous change of measure.*

Appendix A

A Historical Note on Fritz David Carlson (1888–1952)

Fritz David Carlson was born in Vimmerby, where his father, John Vilhelm Carlson, was a house and land holder; his mother was Lovisa Mathilda Carlson. He studied at secondary schools in Vimmerby (1899-1903) and in Linköping (1903-1907) sitting his final high school graduation examination on 4 June 1907. He matriculated as a student at Uppsala University on 5 September 1907. He was awarded his Master of Science degree on 30 May 1911, Licentiate on 14 December 1912, and defended his Ph.D. thesis "On a class of Taylor series" [20] on 26 May 1914 (undertaken with the supervision of Professor Anders Wiman), and became Doctor of Philosophy on 30 May 1914.

As a Liljewalch scholar he visited the University of Göttingen during the period January to August 1916, the University of Berlin and the Technical University in Berlin-Charlottenburg from September to December 1916, and as a Thunsk scholar he also visited Paris in 1920.

He was vicarious teacher at the State School in Vimmerby (1910-11), then auxiliary actuary of the Fire and Life Insurance Company Svea (2 June - 8 September 1911). On 3 June 1914 he became docent in mathematics at Uppsala University ("docent" in the Swedish system means "habilitation" in other countries). During the period from 1914 to 1915 he was a deputy at the Teacher Studies in Uppsala, partly as professor of mathematics at Uppsala University (7 May - 27 May 1920) and then, from 9 July 1920, he was appointed as professor of descriptive geometry at the Royal Institute of Technology (KTH) in Stockholm, a position which he held until 1927.

On 8 August 1923 he married a dentist Marie-Louise Ljungberger (born on 24 June 1894), a daughter of Johan August Ljungberger. Her family ran a bookshop in Uppsala. Marie-Louise died in 1978.

Starting in 1923 he worked for several years as a censor at the student examination. This now historic activity started in 1862 when the universities ceased to have entrance examinations and ended in 1968. The replacement was a flying inspection in which teams of university professors went to the gymnasiums as auditors of the oral examinations and censors when the grades were decided.

Frostman [26] has written the following about the atmosphere of his work as a censor (see also [29], p. 213):

> *In this task he found ample use not only of his experience of pure mathematics but also of his vast knowledge of literature, history, geography and French literature. Many high school teachers, whose teaching he supervised, keep the memory of a demanding censor with a certain stern sense of humour but also of superior comprehension and unfailing judgement.*

From 1927 he became a professor of higher mathematical analysis at the Stockholm University College. Carlson was a real professor in the Swedish tradition of that time. We can quote here Gårding who has written in 1997 on page 212 of [29]:

> *Carlson was one of those who with refined methods continued Mittag-Leffler's effort in the theory of analytic functions. In his daily life he personified the correct professor ...*

and also Kjellberg in 1995 on page 92 of [46]:

> *He was a perfectionist and he could be strict in social intercourse. I think he liked me because I wrote about somewhat old-fashioned things within function theory. In any case he showed me great kindness. Carlson held a very careful oral examination after a student had been awarded a pass in a written examination. It could happen that he started with a candidate in the morning, sent him for lunch, continued for a couple of hours, and then failed him.*

He was a member of the Royal Swedish Academy of Sciences from 1927, the Society of Sciences in Uppsala from 1928, the Royal Physiographic Society of Lund from 1940, and one of the editors of *Acta Mathematica* from 1930. After Carleman's death in 1949 he administrated the Mittag-Leffler's Institute.

Carlson had only three Ph.D. students: H. Rådström (1952), T. Ganelius (1953) and G. Dahlquist (1958).

Carlson died on 28 November 1952 in Stockholm.

His main work focused on the theory of analytic functions. Some of his most well-known contributions are a theorem connected to the Phragmén–Lindelöf principle, a theorem about the zeros of the V-function and several theorems about power series with integer coefficients. In addition to *Carlson's inequalities*, such names as *Carlson's theorem in complex analysis*, *Pólya–Carlson theorem on rational functions* and *the Carlson theorem on Dirichlet series* are well-known in mathematics (see eg. [21] and [22] as well as [34] and [48]).

Carlson's theorem in complex analysis, says that if f is an analytic function satisfying $|f(z)| \leq Ce^{k|z|}$ for $\operatorname{Re} z \geq 0$, where $k < \pi$, and if $f(z) = 0$ for $z = 0, 1, 2, \ldots$, then f is identically zero (cf. [76]).

In a series of papers, Carlson investigated Dirichlet series, and proved in 1922 that if

$$f(z) = \sum_{k=1}^{\infty} a_k k^{-z}$$

is convergent in $\operatorname{Re} z \geq 0$ and bounded in every region $\operatorname{Re} z > \delta > 0$, then, for each $\sigma > 0$,

$$\sum_{k=1}^{\infty} |a_k|^2 k^{-2\sigma} = \lim_{T \to \infty} \frac{1}{2T} \int_{-T}^{T} |f(\sigma + it)|^2 \, dt.$$

Carlson wrote a solid *Textbook of geometry* in two volumes (Gleerup, Lund, 1943, 1947). This resulted from his teaching at KTH, covering first year geometry at the university. He also published a book *Geometry of Space* (Uppsala, 1949).

At a mature age Carlson could remark that (see [29]):

> ... *every mathematician ought to do some work with the Riemann ζ-function.*

Carlson published over thirty papers in mathematics.

See also [80].

This historical note is adopted, with slight modifications, from the web page [63].

Appendix B

A Translation of the Original Article by Carlson from French to English

An Inequality

By

FRITZ CARLSON

We consider the inequality

$$\left(\sum_1^\infty a_n\right)^4 < \pi^2 \sum_1^\infty a_n^2 \sum_1^\infty n^2 a_n^2 \tag{B.1}$$

where a_n are real numbers. It may be regarded as a borderline case of Hölder's inequality

$$\left(\sum_1^\infty a_n\right)^4 \leq \sum_1^\infty a_n^2 \sum_1^\infty n^{2h} a_n^2 \left(\sum_1^\infty n^{-h}\right)^2 = C(h) \sum_1^\infty a_n^2 \sum_1^\infty n^2 a_n^2$$

but here $C(h) \to \infty$ as $h \to 1$, so in this way we cannot conclude the existence of a K such that

$$\left(\sum_1^\infty a_n\right)^4 \leq K \sum_1^\infty a_n^2 \sum_1^\infty n^2 a_n^2. \tag{B.2}$$

Let $f(z) = \sum_1^\infty a_n z^n$. Then

$$f(1)^2 = 2\int_0^1 f(z)f'(z)\,dz < 2\int_{-1}^1 |ff'\,dz| \leq \int_0^{2\pi} |f(e^{i\varphi})f'(e^{i\varphi})|\,d\varphi,$$

$$f(1)^4 < \int_0^{2\pi} |f^2|\,d\varphi \int_0^{2\pi} |f'|^2\,d\varphi = 4\pi^2 \sum_1^\infty a_n^2 \sum_1^\infty n^2 a_n^2.$$

Thus (B.2) holds with $K = 4\pi^2$. To determine the correct value of the constant K, let us consider initially a finite set of numbers a_ν. Suppose that

$$a_\nu \geq 0, \quad \sum_1^N a_\nu = 1, \quad \sum_1^N a_\nu^2 = R$$

and seek the lower bound of

$$S = \sum_1^N \nu^2 a_\nu^2.$$

We will have $a_\nu = \dfrac{\lambda}{\nu^2 + \mu}$, where λ and μ are real. Thus the lower bound of σ in the relation

$$\sum_1^N a_\nu^2 \sum_1^N \nu^2 a_\nu^2 - \sigma \left(\sum_1^N a_\nu \right)^4 = 0$$

is to be deduced from the relation (see the note * on page 190)

$$\sum_1^N \frac{1}{(\nu^2+\mu)^2} \sum_1^N \frac{\nu^2}{(\nu^2+\mu)^2} - \sigma \left(\sum_1^N \frac{1}{\nu^2+\mu} \right)^4 = 0$$

where μ is real. We conclude that the sought value of $1/K$ equals the lower bound of the function ω of z defined by

$$\sum_1^\infty \left(\frac{2z^2}{\nu^2 - z^2} \right)^2 \sum_1^\infty \nu^2 \left(\frac{2z^2}{\nu^2 - z^2} \right)^2 - \omega \left(\sum_1^\infty \frac{2z^2}{\nu^2 - z^2} \right)^4 = 0 \qquad \text{(B.3)}$$

for real and purely imaginary values of z ($z^2 < 1$). We can write ($\pi z = u$)

$$\pi^2 \omega (u \cos u - \sin u)^4 = u^3 (u - \sin u \cos u)(u^2 + u \sin u \cos u - 2 \sin^2 u).$$

It is easily seen that there is $\delta > 0$ such that $\pi^2\omega > 1 + \delta$ for all real values of u. For purely imaginary values of u we always have $\pi^2\omega > 1$ and $\pi^2\omega \to 1$ as u tends to infinity. It follows on the one hand that the relation

$$\left(\sum_1^\infty a_n \right)^4 \leq \pi^2 \sum_1^\infty a_n^2 \sum_1^\infty n^2 a_n^2 \qquad \text{(B.4)}$$

is always valid, on the other hand that to any given $\epsilon > 0$ there correspond numbers a'_ν such that

$$\left(\sum_1^\infty a'_\nu\right)^4 > (\pi^2 - \epsilon)\sum_1^\infty a'^2_\nu \sum_1^\infty \nu^2 a'^2_\nu. \tag{B.5}$$

Thus the correct value of K in (B.2) is π^2.

Let us leave now the relation (B.2) presumedly valid for any sequence a_n. We also have

$$\left(\int_0^\infty \varphi(x)\,dx\right)^4 \leq K\int_0^\infty \varphi^2(x)\,dx \int_0^\infty x^2\varphi^2(x)\,dx. \tag{B.6}$$

Let us apply (B.6) to the function

$$\varphi(x) = e^{-\frac{x}{2}}\sum_0^\infty (-1)^n a_{n+1}\rho^n L_n(x),$$

$L_n(x)$ denoting the Laguerre polynomials and ρ a real number smaller than 1. When $\rho \to 1$ we find

$$\begin{aligned}\left(\sum_1^\infty a_n\right)^4 &\leq \frac{K}{16}\sum_1^\infty a_n^2 \cdot \\ &\quad\cdot \sum_1^\infty\{(6n^2 - 6n + 2)a_n^2 + 8(n-1)^2 a_{n-1}a_n + \\ &\qquad + 2(n-1)(n-2)a_{n-2}a_n\} \\ &< K\sum_1^\infty a_n^2 \sum_1^\infty\left(\left(n - \frac{1}{2}\right)^2 + \frac{3}{16}\right)a_n^2.\end{aligned} \tag{B.7}$$

We conclude from this that the sign of equality can never appear in (B.2) (unless all the a_n vanish), that the inequality (B.1) is valid and that it can be replaced by

$$\left(\sum_1^\infty a_n\right)^4 < \pi^2\sum_1^\infty a_n^2 \sum_1^\infty\left(\left(n - \frac{1}{2}\right)^2 + \frac{3}{16}\right)a_n^2. \tag{B.8}$$

According to (3.1) we have

$$\left(\int_0^\infty \varphi(x)\,dx\right)^4 \leq \pi^2\int_0^\infty \varphi^2(x)\,dx \int_0^\infty x^2\varphi^2(x)\,dx \tag{B.9}$$

and here the sign of equality holds for $\varphi(x) = \dfrac{1}{1+x^2}$.

Inequality (B.1) gives

$$\left(\sum_1^\infty a_n \sum_1^\infty b_n\right)^2 < \frac{\pi^2}{2}\left\{\sum_1^\infty a_n^2 \sum_1^\infty n^2 b_n^2 + \sum_1^\infty b_n^2 \sum_1^\infty n^2 a_n^2\right\} \tag{B.10}$$

$$\left(\int_0^\infty \int_0^\infty \varphi(x)\psi(y)\,dx\,dy\right)^2 \le \frac{\pi^2}{2}\int_0^\infty \int_0^\infty \varphi(x)^2\psi(y)^2(x^2+y^2)\,dx\,dy. \tag{B.11}$$

In (B.11) there will be equality for $\varphi(x) = \psi(x) = \dfrac{1}{1+x^2}$. The integrand is a product $\varphi(x)\psi(y)$; in this respect we remark that there is no relation

$$\left(\int_0^\infty \int_0^\infty F(x,y)\,dx\,dy\right)^2 \le C\int_0^\infty \int_0^\infty F(x,y)^2(x^2+y^2)\,dx\,dy$$

for symmetric functions F.

$$\left(\text{Example} : F = \frac{1}{(x^2+y^2)\{1+|\log(x^2+y^2)|\}^s}\right).$$

Let us finally remark that there is no relation of the form

$$\left(\sum_1^\infty a_n\right)^4 \le C\sum_1^\infty a_n^2 \sum_1^\infty \lambda_n a_n^2, \quad \frac{\lambda_n}{n^2} \to 0.$$

This can be seen by taking $a_n = \rho^n$, $\rho \to 1$.

---

* We may suppose that $R < 1$. The lower bound of S is attained for a set $a_\nu = \alpha_\nu$: $\alpha_1 \ge \alpha_2 \ge \cdots \ge \alpha_N \ge 0$. We necessarily have $\alpha_N > 0$. Indeed, for $a \ge b > 0$, $n > 0$, it is possible to choose $x, y, z > 0$ such that $x+y+z = a+b$, $x^2+y^2+z^2 = a^2+b^2$, $n^2x^2+(n+1)^2y^2+(n+2)^2z^2 < n^2a^2+(n+1)^2b^2$. This shows that the lower bound is a minimum value of S obtained for $a_n = \dfrac{\lambda}{n^2+\mu}$. Let us fix N and vary R; the lower bound of $\sum_1^N a_n^2 \sum_1^N n^2 a_n^2$ for $a_n \ge 0$, $\sum_1^N a_n = 1$ will be equal to the smallest value σ_N of the function σ of μ. This value σ_N is attained for $\mu = \mu_N$, $0 < \mu_N < 3N$

$(N \geq 6)$. If W denotes the lower bound of $\sum_1^N a_n^2 \sum_1^N n^2 a_n^2$ for $a_n \geq 0$, $\sum_1^N a_n = 1$, we will have $\sigma_N \geq W$ and $\sigma_N - W < \epsilon$ for N sufficiently large. Finally one has only to note that σ tends to the function T:

$$\sum_1^\infty \frac{1}{(n^2+\mu)^2} \sum_1^\infty \frac{n^2}{(n^2+\mu)^2} - T\left(\sum_1^\infty \frac{1}{n^2+\mu}\right)^4 = 0,$$

uniformly for $0 \leq \mu \leq 4N$, the interval containing the point $\mu = \mu_N$.

Bibliography

1. M. E. Andersson. Local variants of the Hausdorff-Young inequality. In *Analysis, algebra, and computers in mathematical research (Luleå, 1992)*, volume 156 of *Lecture Notes in Pure and Appl. Math.*, pages 25–34. Dekker, New York, 1994.
2. F. I. Andrianov. "Multi-dimensional analogs of the Carlson inequality and of its generalization". *Izv. Vysš. Učebn. Zaved. Matematika*, 1967(1:56):3–7, 1967. (in Russian).
3. N. Aronszajn and E. Gagliardo. Interpolation spaces and interpolation methods. *Ann. Mat. Pura Appl. (4)*, 68:51–117, 1965.
4. S. Barza. *Weighted Multidimensional Integral Inequalities and Applications.* PhD thesis, Luleå Institute of Technology, 1999.
5. S. Barza, V. Burenkov, J. Pečarić, and L.-E. Persson. Sharp multidimensional multiplicative inequalities for weighted L_p spaces with homogeneous weights. *Math. Inequal. Appl.*, 1(1):53–67, 1998.
6. S. Barza, J. Pečarić, and L.-E. Persson. Carlson type inequalities. *J. Inequal. Appl.*, 2(2):121–135, 1998.
7. S. Barza and E. C. Popa. Inequalities related with Carlson's inequality. *Tamkang J. Math.*, 29(1):59–64, 1998.
8. E. F. Beckenbach and R. Bellman. *Inequalities.* Ergebnisse der Mathematik und ihrer Grenzgebiete, N. F., Bd. 30. Springer-Verlag, Berlin, 1961.
9. R. Bellman. An integral inequality. *Duke Math. J.*, 10:547–550, 1943.
10. E. I. Berezhnoĭ. Interpolation of linear and compact operators in the spaces $\varphi(X_0, X_1)$. In *Qualitative and approximate methods for the investigation of operator equations (Russian)*, pages 19–29, 165. Yaroslav. Gos. Univ., Yaroslavl′, 1980.
11. J. Bergh. An optimal reconstruction of sampled signals. *J. Math. Anal. Appl.*, 115(2):574–577, 1986.
12. J. Bergh and J. Löfström. *Interpolation spaces. An introduction.* Springer-Verlag, Berlin, 1976. Grundlehren der Mathematischen Wissenschaften, No. 223.
13. J. I. Bertolo and D. L. Fernandez. On the connection between the real and the complex interpolation method for several Banach spaces. *Rend. Sem. Mat.*

Univ. Padova, 66:193–209, 1982.

14. J. I. Bertolo and D. L. Fernandez. A multidimensional version of the Carlson inequality. *J. Math. Anal. Appl.*, 100(1):302–306, 1984.
15. A. Beurling. Sur les intégrales de Fourier absolument convergentes et leur application à une transformation fonctionelle. *C. R. Neuvième Congrès Math. Scandinaves 1938*, pages 345–366, 1939.
16. P. Brenner and V. Thomée. On rational approximations of semigroups. *SIAM J. Numer. Anal.*, 16(4):683–694, 1979.
17. Y. A. Brudnyĭ and N. Y. Krugljak. *Real Interpolation Functors*. VINITI N 2620-81, Yaroslavl, 1981. in Russian.
18. Y. A. Brudnyĭ and N. Y. Krugljak. *Interpolation functors and interpolation spaces. Vol. I*, volume 47 of *North-Holland Mathematical Library*. North-Holland Publishing Co., Amsterdam, 1991. Translated from the Russian by Natalie Wadhwa, With a preface by Jaak Peetre.
19. A.-P. Calderón. Intermediate spaces and interpolation, the complex method. *Studia Math.*, 24:113–190, 1964.
20. F. Carlson. *Sur une classe de séries de Taylor*. PhD thesis, Uppsala University, 1914.
21. F. Carlson. Über die Nullstellen der Dirichlet'schen Reihen und der Riemann'schen ζ-funktion. *Ark. Mat. Astr. Fys.*, 15(20), 1921.
22. F. Carlson. Contributions à la théorie des séries de Dirichlet. *Ark. Mat. Astr. Fys.*, 16(18), 1922.
23. F. Carlson. Une inégalité. *Ark. Mat. Astr. Fysik*, 25B(1), 1935.
24. W. B. Caton. A class of inequalities. *Duke Math. J.*, 6:442–461, 1940.
25. D. L. Fernandez and J. B. Garcia. Interpolation of Orlicz-valued function spaces and U.M.D. property. *Studia Math.*, 99(1):23–40, 1991.
26. O. Frostman. Fritz Carlson in memoriam. *Acta Math.*, 90:ix–xii, 1953.
27. R. M. Gabriel. An extension of an inequality due to Carlson. *J. London Math. Soc.*, 12:130–132, 1937.
28. J. B. Garcia. Interpolation of weighted Orlicz-valued function spaces. *Anal. Math.*, 19(3):191–215, 1993.
29. L. Gårding. *Matematik och matematiker*. Lund University Press, Lund, 1994. Matematiken i Sverige före 1950. [Mathematics in Sweden before 1950].
30. J. Gustavsson. On interpolation of weighted L^p-spaces and Ovchinnikov's theorem. *Studia Math.*, 72(3):237–251, 1982.
31. J. Gustavsson and J. Peetre. Interpolation of Orlicz spaces. *Studia Math.*, 60(1):33–59, 1977.
32. G. H. Hardy. A note on two inequalities. *J. London Math. Soc.*, 11:167–170, 1936.
33. G. H. Hardy and J. E. Littlewood. *Inequalities*. Gosudarstv. Izdat. Inostr. Lit., Moscow, 1948.
34. G. H. Hardy, J. E. Littlewood, and G. Pólya. *Inequalities*. Cambridge University Press, 1952. 2nd ed.
35. G. W. Hedstrom. Norms of powers of absolutely convergent Fourier series in several variables. *Michigan Math. J.*, 14:493–495, 1967.
36. K. Hu. "On the Nagy–Carlson type inequalities". *J. Jiangxi Norm. Univ.*

Nat. Sci. Ed., 17(2):89–90, 1993. (in Chinese).
37. S. Janson. Minimal and maximal methods of interpolation. *J. Funct. Anal.*, 44(1):50–73, 1981.
38. A. Kamaly. On certain inequalities for Fourier coefficients. *SUT J. Math.*, 34(2):91–103, 1998.
39. A. Kamaly. *Type-Inequality Problems in Fourier Analysis*. PhD thesis, Royal Institute of Technology, 1998.
40. A. Kamaly. Fritz Carlson's inequality and its application. *Math. Scand.*, 86(1):100–108, 2000.
41. A. Y. Karlovich and L. Maligranda. On the interpolation constant for Orlicz spaces. *Proc. Amer. Math. Soc.*, 129(9):2727–2739 (electronic), 2001.
42. B. Kjefsjö. Problem in Elementa. *Elementa*, 54(2), 1971.
43. B. Kjellberg. Ein Momentenproblem. *Ark. Mat. Astr. Fys.*, 29A(2):1–33, 1943.
44. B. Kjellberg. On some inequalities. In *C. R. Dixième Congrès Math. Scandinaves 1946*, pages 333–340. Jul. Gjellerups Forlag, Copenhagen, 1947.
45. B. Kjellberg. A note on an inequality. *Ark. Mat.*, 3:293–294, 1956.
46. B. Kjellberg. Mathematicians in Uppsala—some recollections. In *Festschrift in honour of Lennart Carleson and Yngve Domar (Uppsala, 1993)*, volume 58 of *Acta Univ. Upsaliensis Skr. Uppsala Univ. C Organ. Hist.*, pages 87–95. Uppsala Univ., Uppsala, 1995.
47. N. Y. Kruglyak, L. Maligranda, and L.-E. Persson. A Carlson type inequality with blocks and interpolation. *Studia Math.*, 104(2):161–180, 1993.
48. L. Larsson. *Carlson Type Inequalities and their Applications*. PhD thesis, Uppsala University, 2003.
49. L. Larsson. A multidimensional extension of a Carlson type inequality. *Indian J. Pure Appl. Math.*, 34(6):941–946, 2003.
50. L. Larsson. A new Carlson type inequality. *Math. Inequal. Appl.*, 6(1):55–79, 2003.
51. L. Larsson. Carlson type inequalities and embeddings of interpolation spaces. *Proc. Amer. Math. Soc.*, 132(8):2351–2356 (electronic), 2004.
52. L. Larsson, Z. Páles, and L.-E. Persson. Carlson type inequalities for finite sums and integrals on bounded intervals. *Bull. Austral. Math. Soc.*, 71(2):275–284, 2005.
53. L. Larsson, J. Pečarić, and L.-E. Persson. An extension of the Landau and Levin-Stečkin inequalities. *Acta Sci. Math. (Szeged)*, 70(1-2):25–34, 2004.
54. N. Leblanc. Sur la réciproque de l'inégalité de Carlson. *C. R. Acad. Sci. Paris Sér. A-B*, 267:A332–A334, 1968.
55. V. I. Levin. "Some inequalities between series". *Mat. Sb.*, 45:341–345, 1938. (in Russian).
56. V. I. Levin. "Exact constants in inequalities of the Carlson type". *Doklady Akad. Nauk SSSR (N.S.)*, 59:635–638, 1948. (in Russian).
57. V. I. Levin and E. K. Godunova. "A generalization of Carlson's inequality". *Mat. Sb. (N.S.)*, 67 (109):643–646, 1965. (in Russian).
58. V. I. Levin and S. B. Stečkin. Inequalities. *Amer. Math. Soc. Transl. (2)*, 14:1–29, 1960.

59. G. J. Lozanovskiĭ. A remark on a certain interpolation theorem of Calderón. *Funkcional. Anal. i Priložen.*, 6(4):89–90, 1972.
60. L. Maligranda. Calderón–Lozanovskiĭ spaces and interpolation of operators. *Semesterbericht Funtionalanalysis*, 8:83–92, 1985.
61. L. Maligranda. *Orlicz Spaces and Interpolation*, volume 5 of *Seminários de Matemática*. Univ. Estadual de Campinas, Campinas SP, Brazil, 1989.
62. L. Maligranda. Why Hölder's inequality should be called Rogers' inequality. *Math. Inequal. Appl.*, 1(1):69–83, 1998.
63. L. Maligranda. Fritz Carlson. `http://www-history.mcs.st-andrews.ac.uk/Mathematicians/Carlson.html`, 2005.
64. L. Maligranda and L.-E. Persson. Real interpolation between weighted L^p and Lorentz spaces. *Bull. Polish Acad. Sci. Math.*, 35(11-12):765–778, 1987.
65. V. G. Maz'ja. *Sobolev spaces.* Springer Series in Soviet Mathematics. Springer-Verlag, Berlin, 1985. Translated from the Russian by T. O. Shaposhnikova.
66. D. S. Mitrinović. *Analytic inequalities.* In cooperation with P. M. Vasić. Die Grundlehren der mathematischen Wisenschaften, Band 1965. Springer-Verlag, New York, 1970.
67. D. S. Mitrinović, J. E. Pečarić, and A. M. Fink. *Classical and new inequalities in analysis*, volume 61 of *Mathematics and its Applications (East European Series)*. Kluwer Academic Publishers Group, Dordrecht, 1993.
68. P. Nilsson. Interpolation of Banach lattices. *Studia Math.*, 82(2):135–154, 1985.
69. K. I. Oskolkov. Approximation properties of summable functions on sets of full measure. *Math. USSR Sbornik*, 32(4):489–517, 1977. in Russian.
70. V. I. Ovchinnikov. Interpolation theorems following from Grothendieck's inequality. *Funtional Anal. Appl.*, 10:287–294, 1977.
71. V. I. Ovchinnikov. The method of orbits in interpolation theory. *Math. Rep.*, 1(2):i–x and 349–515, 1984.
72. J. Peetre. Sur le nombre de paramétres dans la définition de certains espaces d'interpolation. *Ricerche Mat.*, 12:248–261, 1963.
73. J. Peetre. Sur l'utilisation des suites inconditionellement sommables dans la théorie des espaces d'interpolation. *Rend. Sem. Mat. Univ. Padova*, 46:173–190, 1971.
74. G. M. Pigolkin. "Multidimensional inequalities of Carlson type and others connected with them". *Sibirsk. Mat. Ž.*, 11:151–160, 1970. (in Russian).
75. R. Redheffer. Integral inequalities with boundary terms. In *Inequalities, II (Proc. Second Sympos., U.S. Air Force Acad., Colo., 1967)*, pages 261–291. Academic Press, New York, 1970.
76. M. Reed and B. Simon. *Methods of modern mathematical physics. IV. Analysis of operators.* Academic Press [Harcourt Brace Jovanovich Publishers], New York, 1978.
77. S. Roudenko. Matrix-weighted Besov spaces. *Trans. Amer. Math. Soc.*, 355(1):273–314 (electronic), 2003.
78. V. A. Shestakov. "Transformations of Banach lattices and interpolation of

linear operators". *Bull. Polish Sci. Math.*, 29:569–577, 1981. (in Russian).

79. P. Sjölin. A remark on the Hausdorff-Young inequality. *Proc. Amer. Math. Soc.*, 123(10):3085–3088, 1995.
80. *Svenskt Biografiskt Lexikon*, volume 7. Albert Bonniers Förlag, 1927.
81. B. Sz. Nagy. "On Carlson's and related inequalities". *Mat. Fiz. Lapok*, 48:162–175, 1941. (in Hungarian).
82. B. Sz. Nagy. Über Integralungleichungen zwischen einer Funktion und ihrer Ableitung. *Acta Univ. Szeged. Sect. Sci. Math.*, 10:64–74, 1941.
83. S. Treil and A. Volberg. Completely regular multivariate stationary processes and the Muckenhoupt condition. *Pacific J. Math.*, 190(2):361–382, 1999.
84. G.-S. Yang and J.-C. Fang. Some generalizations of Carlson's inequalities. *Indian J. Pure Appl. Math.*, 30(10):1031–1040, 1999.

Index